ELITES IN CONFLICT

Elites in Conflict

The Antebellum Clash over the Dudley Observatory

Mary Ann James

Rutgers University Press
New Brunswick and London

Publication of this book has been aided by a grant from the Dudley Observatory.

Library of Congress Cataloging-in-Publication Data

James, Mary Ann, 1940–
Elites in conflict.

Bibliography; p.
Includes index.
1. Dudley Observatory—History. 2. Astronomical observatories—New York (State)—Albany—History. 3. Science—New York (State)—Albany—History. I. Title.
QB82.U62A435 1987 522′.1974743 87-4305
ISBN 0-8135-1245-X

British Cataloging-in-Publication information available.

For Henry

Contents

Illustrations

Acknowledgments

By the time anyone has completed a book, the list of those who contributed time, effort, and knowledge is daunting. To mention everyone is impossible. Each of us, to paraphrase Thomas Wolfe, is the sum of all we cannot count. My own debts are extensive. Although I cannot enumerate the many colleagues, manuscript librarians, and archivists—some whose names I will never know—who cheerfully offered their help, special thanks have to be given to the staff at the Houghton Library and the Archives at Harvard University, who were always unfailing in their helpful professionalism. I must also thank the manuscript librarians at the Albany Institute of History and Art, the Cincinnati Historical Society, the Franklin Institute, the Library of Congress, the Huntington Library, and the Greenwich Observatory Archives. The endnotes serve as a testimonial to the helpfulness of the staffs and librarians at these and other institutions.

Some intellectual debts are obvious. Several scholars sacrificed time from their own work to offer valuable criticism. Dr. Nathan Reingold, Senior Historian at the Smithsonian Institution and former editor of the Joseph Henry Papers, was extraordinarily generous in reading the manuscript and providing thoughtful comments. Dr. Reingold's extensive knowledge of the nineteenth-century scientific community was invaluable in establishing connections as well as in pointing out errors. His observations were both thought-provoking and stimulating. The manuscript also benefited enormously from the insightful and candid comments offered by Dr. Marc Rothenberg, acting editor of the Joseph Henry Papers, and Dr. John Lankford, Professor of the Social History of Modern Science at the University of Missouri at Columbia. Professor Owen Gingerich, Smithsonian Professor of Astrophysics and the History of Science at Harvard University, offered helpful corrections on an earlier version of the manuscript.

Other intellectual debts are impossible to calculate or even to express in a meaningful way. Professors Albert Van Helden and Thomas Haskell of the history department at Rice University were unfailing in their patience, continuing interest, and support. Each contributed immeasurably. One friend deserves my special thanks. From the beginning, Dr. Ruth Whiteside has been a never-ending source of interesting ideas and fresh perspectives. Only she and I understand how valuable her contributions have always been. Professor Jan Ludwig of Union College must also be mentioned for providing encouragement and a willing ear. My colleagues at Rutgers University cheerfully took time from busy schedules and their own research to give me the kind of close reading and serious criticism that only comes from good friends. Thanks are also due to Kathy Fluss, my intrepid research assistant during this past year, and to Karen Reeds and Marilyn Campbell of Rutgers University Press. Everyone has offered suggestions along the way, not all of which I have taken. The errors, therefore, belong only to me.

Finally, I must thank my family. I don't know how to begin to acknowledge my debt to my parents. I assume they know how much they both mean to me. The joyful noise that is part of living with three wonderful children—Michael, Matthew, and Kathryn—has continually reminded me to keep a sense of humor in my scholarly pursuits. My greatest debt is to my husband, who deserves the "above and beyond" award if anyone does. Henry lived with this project from its inception and learned more about the conflict in Albany and nineteenth-century scientists than I'm sure he ever wanted to know. His faith and confidence were critical factors in the manuscript's completion. The Pollock Award for Research in the History of Astronomy provided generous financial support for the final stages of research and writing.

Cast of Characters

George Biddle Airy—the Astronomer Royal at Greenwich Observatory.

Dr. James H. Armsby—local physician; principal organizer of the law school, medical school, hospital, observatory, and veterans' hospital in Albany.

Alexander Dallas Bache—one of the most influential members of the scientific community in the United States; superintendent of the United States Coast Survey from 1843 until his death in 1867.

William Cranch Bond and George Phillips Bond—the first director of the Harvard Observatory and his son, who succeeded him.

Howell Cobb—secretary of the Treasury under President Buchanan; Bache's superior in the Treasury Department; formerly most powerful congressman in the House of Representatives.

Erastus Corning—wealthy iron merchant; president of the first New York Central Railroad; known as "the Railroad King"; former mayor of Albany; influential congressman.

Jefferson Davis—senator from Mississippi during the time of the Dudley Observatory dispute; close friend to Bache.

Mrs. Blandina Dudley—widow of late U.S. Senator Charles Dudley; known for her philanthropy in Albany.

Oliver Wolcott Gibbs—European-educated chemist who introduced German methods of laboratory research to the United States; became Rumford Professor at Harvard in 1863.

Benjamin Apthorp Gould, Jr.—first director of the Dudley Observatory; would later make his reputation as director of the National Observatory of Argentina.

James Hall—member of the geological survey of New York authorized by Governor Marcy in 1836; named state paleontologist in 1843.

Ira Harris—justice of the Supreme Court of the State of New York; became United States senator from New York in 1861.

Joseph Henry—respected scientist from Albany; first secretary of the Smithsonian Institution; involved in controversy with Samuel Morse and disputes over the proper use of the Smithson endowment.

Junius Hillyer—solicitor of the Treasury Department during President Buchanan's administration.

Rutger B. Miller—Mrs. Dudley's nephew.

Ormsby Macknight Mitchel—director of the Cincinnati Observatory; known for his spellbinding lectures on astronomy.

Thomas Worth Olcott—president of the Mechanics' and Farmers' Bank in Albany; respected member of the "Albany Regency"; president of the board of trustees of the Dudley Observatory during the controversy.

Benjamin Peirce—considered to be the foremost mathematician in the United States; Perkins Professor or Mathematics and Astronomy at Harvard University.

Dr. Christian Heinrich Friedrich Peters—European astronomer who emigrated to the United States in 1848.

John Van Schaick Lansing Pruyn—member of the New York Board of Regents; respected member of Albany cultural elite.

General Robert Hewson Pruyn—speaker of the New York Assembly; member of the board of trustees; later appointed as the second United States minister to Japan by President Lincoln in 1861.

John Spencer—district attorney of the Northern District of New York.

Stephen Van Rensselaer III—"the Old Patroon"; the first great patron of science in the United States; established Rensselaer Polytechnic Institute in Troy, New York.

General Stephen Van Rensselaer IV—"the Young Patroon"; president of the board of trustees of the Dudley Observatory when it was first organized.

John Wilder—member of board of trustees of the Dudley Observatory; founder and president of the board of trustees of Rochester University.

ELITES IN CONFLICT

CHAPTER 1

Introduction

Scientific institutions tend to be rather solemn and serious places. We don't normally expect them to erupt in dramatic confrontations or public turmoil. Not only are such scenes embarrassing to all concerned, but they seem somehow contrary to the "image" of science. Today, however begrudgingly we defer to the specialist, most of us assume that if problems arise within scientific institutions the "experts" will sort them out in more appropriate and less public arenas. At the middle of the nineteenth century, emerging professionals were even more reticent than their present-day counterparts to air professional linen in public. The risk of losing public support for newly established institutions or drawing unwanted attention to claims of professional authority was too great. This understandable caution makes the flamboyant quarrel between eminent antebellum scientists and patrons of science over an Albany observatory in the 1850s intriguing on many levels.

The conflict, which was likened to a lawsuit fought out of court, confounded contemporaries because it was taken to such potentially damaging extremes by all concerned. Astronomer Simon Newcomb described the quarrel as more bitter than any contest he had ever seen in American politics.[1] Although the battle was waged with words, it was as strenuously contested as any physical confrontation. Contemporary descriptions inevitably invoked images drawn from the language of hand-to-hand clashes between bitter adversaries. Participants in the quarrel launched themselves repeatedly into print. Barrages of statements, pamphlets, and rebuttals were aimed like cannonballs at the opposing forces by each side. Because those who took part were all men with substantial reputations, escalating events provided journalists across the nation with copy for months. Long after the dust settled, historians used the story to draw wider generalizations about antebellum attitudes toward science and the process of professionalization in the nineteenth century.

What was all the uproar about? In the 1850s institutional roles for scientists were in flux, expanding steadily from college teaching toward the more modern sense of "specialist." During this transitional time, the quarrel at the Dudley Observatory arose out of conflicting perceptions of personal responsibility, accountability, and authority within the institution, as interpreted by two respected nineteenth-century elites—the scientists and their antebellum patrons.[2] On the face of it, the controversy was over the right of the scientists to make decisions about spending money, hiring staff, committing to projects, or summarily closing the observatory without consulting the institution's non-scientist trustees. The scientists argued that they had absolute autonomy over the institution to which they had attached their reputations. The trustees insisted that they were publicly responsible for making scrupulously certain that outlays were justified and that the institution continued to progress. Individual reputations of powerful men on both sides were attached to the observatory; as a result, the conflict was both bitter and personal.

Although the quarrel was acrimonious and the animosities long-lasting, the participants were all men of good intentions. On one side, dedicated scientists were committed to an emerging professional ethnic. On the other, men of sound reputation in business and politics were committed to an ethic of responsible stewardship. The two value systems are not necessarily opposed but, in this instance, brought two influential elites into an open conflict that raised other sensitive issues on the eve of the Civil War. These two elites had come together eagerly to establish an outstanding research institution in the United States, one equipped to compete with the best in Europe. Instead, they found themselves inexorably drawn into an altercation that spread far beyond the doors of the scientific institution in which it began, spilling out into the corridors of the state legislature, spreading into the halls of Congress, and finally washing up into the offices of members of President Buchanan's Cabinet.

The scientists in this unpleasant and increasingly undignified quarrel included several of the most eminent in the nation. Each would be remembered for his efforts to elevate the standards of science in the United States to European levels of rigor. Each would be one of the founding members of the National Academy of Sciences. Three of the four were already associated with the most prestigious American scientific institutions of the 1850s. Alexander Dallas Bache was the politically astute superintendent of the United States Coast Survey, the most important source of federal funding for science in antebellum America. Joseph Henry, a respected physicist from Albany by way of Princeton University, was the first secretary of the Smithsonian Institution. Benjamin Peirce, the most famous American mathematician of his day, was Perkins Professor of Mathematics and Astronomy at Harvard University and would succeed Bache as superintendent of the Coast Survey. The astronomer at the center of the stormy controversy, Benjamin Apthrop Gould, Jr., was one of the first Americans to receive a doctorate in science from a European university and the first to be offered the directorship of a major European observatory. Gould would eventually become the director of the National Observatory of Argentina. These four impressive professionals made up the Scientific Council of the Dudley Observatory in Albany, New York.

Their opponents were men of wealth, power, and influence in both business and national politics. Thomas Worth Olcott was a banker of national reputation, the architect of the first central banking system in the United States. Olcott acquired an early reputation as the astute financier of the "Albany Regency," which controlled New York politics for decades. In 1863 President Lincoln would ask the Albany banker to become the first comptroller of the currency. Ira Harris, a justice on the New York State Supreme Court during the conflict, would soon be named United States senator from New York. Other powerful men, such as Stephen Van Rensselaer, known as the "last patroon" and heir to one of the largest estates in American history, made up the board of trustees of the Dudley Observatory.[3]

These two eminent elites shared a common goal. They planned to establish a scientific institution in Albany that would achieve international acclaim. They ended by coming as close to blows as was conceivable for those who considered themselves to be gentlemen. In many ways only Gilbert and Sullivan could do justice to the more melodramatic aspect of the conflict. As the quarrel escalated, some of the events almost seem drawn from the scenes of a classic farce. But events that seem humorous in retrospect were devastating when experienced personally, even more so when humiliations were well publicized. Because the embattled participants on both sides of the quarrel were powerful, many with names that extended far beyond Albany, the heated dispute attracted a mortifying amount of attention at a time when no gentleman cared to have his name or reputation bandied about in the press.

The arguments put forth so vehemently in the rancorous quarrel offer insights into basic assumptions governing the attitudes and actions of scientists and their patrons at the middle of the nineteenth century. Although the dispute has been cast in the mold of "scientist versus layman" and described as an example of the opposition of antebellum Americans to the elitist values of modern science, the conflict is far more complex.[4] This was not a rejection of scientific elitism by Jacksonian Philistines. Both sides in this quarrel were elitists. Nor was anyone arguing against rigorous science of the highest professional standards. On the contrary, everyone involved was in favor of the professional ideology espoused by the scientists. Instead, the perceptual differences obvious in this argument reflect a basic misalignment in the standards by which two influential groups made decisions in a difficult situation. Perhaps the best analogy for their differences is that of the Gestalt choice between an urn or two faces. The conflict in Albany escalated because neither side acknowledged the validity of their opponents' perceptions and standards of judgment.

The standards the trustees brought to the quarrel were based on the serious sense of fiduciary responsibility that prevailed among respected members of antebellum commercial communities during the second quarter of the nineteenth century. In a time of rapidly accelerating social and economic change, elites in cities like Boston, Philadelphia, and Albany were committed to rigorous standards of stewardship nurtured by the Christian denominations. This antebellum concept of stewardship served positively in creating and sustaining many kinds of institutions—charitable, educational, financial, and scientific.

With wealth came responsibilities. Few questioned the right of any man to amass riches through his personal efforts, but many felt they had every right to question the uses to which he might put that wealth. Before the excesses of the Gilded Age, most Americans did not see wealth as an end in itself but as a means to an end. Great fortunes should be used to benefit the community, not squandered in self-indulgent excesses. An ostentatious lifestyle was still frowned upon, with Manhattan perhaps serving as the rule-proving exception. Prevailing middle-class values were even prejudiced against bequeathing large estates to one's own children. Instead, a respected entrepreneur was expected to devote himself to good works that extended beyond the countinghouse and beyond his own enjoyments. Given the antebellum inhibitions on conspicuous consumption, voluntary benevolence offered a means for achieving public recognition and personal gratification in a manner that was both socially approved and self-satisfying. Philanthropy was the final self-pleasurable fruit of complacent success, fully enjoyed by antebellum Americans who amassed or inherited great fortunes.[5]

Philanthropic impulses also conferred their own kind of power. Although foreign observers were struck by the egalitarian aspects of American life, established urban communities in antebellum America were far more stratified than surface appearances indicated. Control over local voluntary associations and cultural institutions gave a quiet urban elite vast influence during a period known for its rhetorical emphasis upon the common man. In recognition of this social reality, self-assured entrepreneurs shifted away from direct control of municipal politics to more subtle but equally effective positions of social power. Boards of trustees of private schools, orphanages, hospitals, rural cemetery associations, and newly-founded scientific institutions included the most powerful members of the urban elite in cities like Boston and Albany.[6]

Responsibilities were taken seriously within these social and cultural institutions, just as seriously as they were assumed to be in banking, insurance, shipping, and commerce. Fiduciary responsibility involved personal honor. The commercial world of antebellum America was not the impersonal corporate one we know today. A successful businessman was not anonymous. His working world was small, heavily dependent upon his reputation and the amount of trust others placed in his character. Standards were rigorous; behavior had to be above reproach, for judgments were far-reaching. The trustees of the Dudley Observatory, as well as some of their major local opponents, were successful and respected members of this commercial elite. They approached their philanthropic commitments with the same values and attitudes they brought to other undertakings bearing their names and carrying their reputations.

Antebellum businessmen of this respected class assumed that a man's word bound his honor, no matter what. Christianity also demanded that men fulfill their obligations. Under the standards of this ethic of stewardship, the greatest test of character came at moments of personal difficulty. A man who committed his name to care for someone else's property was expected to sacrifice his own interests rather than endanger that which had been entrusted to him. If a businessman lost his wealth and declared bankruptcy, his Christian duty demanded that he repay every penny owed as soon as he regained any capital, despite the legal protection provided by the bankruptcy laws. Trust could not be betrayed; to do so was to sin against both religious principles and society. To men of property afloat in the orgy of speculation and financial upheaval that accompanied economic growth between 1830 and 1850, the sacredness of personal obligations was tied closely to morality, to religion, and to the preservation of political stability. Fiscal virtue reflected personal virtue and provided the foundation for the republic. Obviously, fiscal irresponsibility was one of the most serious charges that could be made against an antebellum businessman. In the days when rampant state legislatures were prone to repudiate debts, irresponsibility in matters of trust threatened the entire fabric of economic order. This antebellum emphasis on high personal standards of fiduciary responsibility played an important role in the trustees' assessment of the actions and character of the director of the Dudley Observatory during the conflict.[7]

The scientists involved in the quarrel were themselves a remarkable group, one focused on excellence. Their struggle for institutional change before the Civil War on so many fronts at once—the Coast Survey, the Smithsonian, the American Association for the Advancement of Science, the attempt to establish a national university, the Dudley Observatory, the postbellum establishment of the National Academy of Sciences—is itself extraordinary. The observatory in Albany briefly meshed with their aspirations for American science. Although the Dudley Observatory never became a major institution, for a short time it appeared to have the potential for true greatness, an American Greenwich on the Hudson in the making. The possibilities envisioned for the observatory's future induced three of the most influential American scientists of the time to use all means available to maintain their control of the institution.

The strategic maneuverings that followed as the quarrel developed offer insights into the intricate interweaving of American politics and science at the highest levels right before the Civil War. In the course of the quarrel, Alexander Dallas Bache, one of the ablest scientist-politicians of the nineteenth century, moved deftly through the minefields of the Treasury Department and Congress, played local politics in Albany, balanced a district attorney against the solicitor of the treasury, and brought the weighty influence of Senator Jefferson Davis and powerful congressmen like Erastus Corning to bear in his own behalf. In his attempt to maintain control of the observatory, Bache's arguments invoked a broad definition of the right of the federal government to take control of private property, ideas by no means accepted in 1858 and 1859 but ironically supported in this instance by Jefferson Davis. Because these two influential elites battled each other with all the political influence they could muster on state and federal levels just before the Civil War, the conflict over the Dudley Observatory in Albany fits into a context far broader than the professionalization of science.

Most scholarship has assumed that the observatory foundered because Americans were not ready or willing to support basic science until the economic and social restructuring that followed the Civil War generated an increasing acceptance of professional values. Nevertheless, the conflict in Albany reveals a far more general acceptance of "professional" science than antebellum rhetoric on the problems of professionalization—taken from the letters of antebellum scientists—would indicate. Although the early history of the Dudley Observatory has been used as an example of the lack of support for basic science in antebellum America, the facts of the tale lead to the opposite conclusion.

In any culture some values favor, others inhibit the development of science. Antebellum scientists in cities like Boston, Philadelphia, and Albany benefited from a confluence of attitudes in the second quarter of the nineteenth century. An increasing recognition of the importance of science combined with a strong sense of personal stewardship and civic responsibility to establish and support new patterns for institutional growth. During the

1840s and 1850s, American philanthropists shifted their benevolence away from moral instruction and reform toward intellectual inquiry and scientific research. Of all the sciences, astronomy was the most appreciated by antebellum Americans, in folklore, poetry, and the popular press. Newspapers throughout the United States carried more articles on astronomy than on any other science during the period. Given this widespread interest, it is not surprising that astronomy was the first science to receive extensive support from American philanthropists. Donations offered to establish observatories in Boston, Cincinnati, Detroit, and Albany as well as in other smaller communities in the 1840s and 1850s sprang from many motives, ranging from simple curiosity and local boosterism to a sincere desire to advance knowledge. One theme was constant, however. Each of these observatories reflected a desire to attract international recognition for American achievements. Cultural nationalism was strong in antebellum America.[8]

As Emerson noted, antebellum Americans wanted an authentically American culture, but they also wanted the merit of that culture to be acknowledged abroad. Science was part of these antebellum aspirations. A significant scientific institution in the United States promised increasing European recognition. As a result, civic-minded men with money to spare supported a number of scientific institutions that promised to lift the nation to cultural equality with Europe. American scientists were themselves as desirous of European recognition as local boosters. Many believed that American scientific contributions were either denigrated or ignored by patronizing Europeans. The scientists involved in the Albany quarrel were intent throughout their careers on raising the standards of professional education and scientific research in the United States to achieve international scholarly respect. Antebellum patrons of science supported those aims in a tangible way, with dollars from their own pockets. Despite the reputed American penchant for practicality, antebellum philanthropists, including the trustees of the Dudley Observatory, set a remarkable standard of generosity. Hostility to scientific research in antebellum America has been overestimated by those who concentrate on the rhetoric of professionalizing scientists rather than the reality of actual support.[9]

Patronage, both governmental and individual, offers a strategic perspective on the dynamics of institutional development in science. Private philanthropy has played a vital role in the formation of American institutions. Although the great period of growth in American scientific institutions occurred after the Civil War, the motivations that powered that surge toward institutional specialization are clearly discernible in the second quarter of the nineteenth century in both scientists and patrons of science. The same values reflected in the foundation of new universities like Johns Hopkins and Stanford and observatories like Lick and Yerkes can be observed in American philanthropy before the Civil War. The Dudley Observatory of Albany provides a powerful example of both the confident framing of great plans for the future during the antebellum period and of the studious concentration of resources necessary in making the great institutional leap that came after the Civil War.[10]

The rancorous conflict over the observatory also provides a specific historical perspective in which the intentions of both scientists and patrons of science can be evaluated before that great period of institutional growth. The Albany quarrel offers a field of vision in which the changing boundaries of authority and accountability within scientific institutions can be brought into clearer focus. The parameters of authority and responsibility within antebellum institutions were not clearly defined for either scientist or patron. Although expanding from the position of teacher, the scientist's role as authoritative expert was not yet established. Power within antebellum institutions was firmly held by laymen. Lacking the influence enjoyed by modern scientists, antebellum professionals who hoped to establish long-term institutions had to proceed carefully. Persuading individual patrons to underwrite esoteric and often expensive research was a delicate task. Although Americans were generous and enthusiastic in their financial commitments to antebellum science, the task facing emerging professionals was to channel those financial resources in the correct direction, one that would be meaningful for science in the United States.

If the nineteenth-century scientist were to rise above the status of mere employee (often his role within the antebellum college) to shape an institution to his own research needs, he had to persuade the nonspecialists in charge to defer willingly to his leadership. There were a few American scientists who successfully managed antebellum institutions recognized for the quality of their scientific research—Alexander Dallas Bache at the Coast Survey, Joseph Henry at the Smithsonian, William and George Bond at the Harvard Observatory. James Hall had an equally successful career as state paleontologist for New York. Each of these scientists was skillful in mediating between the providers of funds and his professional needs. Each was trusted. Each acknowledged his own accountability. Each was aware of the need for political sensibility in his particular situation. Except for the Bonds at Harvard Observatory, these scientists established their roles in institutions supported either by the state or by the federal government. Harvard, of course, was itself an atypical situation. The Boston elite had an exceptionally strong philanthropic commitment to the university, and community support for science was equally consistent. The role of the scientist-entrepreneur in an institution completely dependent upon private philanthropy, an institution not associated with either the government or with Harvard University, was less defined and more open to interpretation. The Dudley Observatory was such as institution; its director would chart less-travelled ground. His success would depend on his ability to maintain support for his undertaking.

Obviously, the participants in any kind of voluntary action share similar goals. Of equal importance is their shared sense of mutual trust. Even if goals are shared by all, few will commit themselves fully in energy much less in reputation or money without full confidence in the character and integrity of the person in charge. Antebellum Americans did not have the recognized certificates of competency—graduate degrees—on which modern society relies. No American university granted a doctorate until 1861. How was trust established? Without credentials, one had to rely on personal experience, recommendation, and judgment on character. Once trust was given, however, how could performance be guaranteed? The formalized mechanisms of accountability, especially in newly established institutions, had to be carefully worked out in the antebellum social context with few models for guidance. Today, it is obvious that the director of any research institution must constantly educate his patrons whether they are congressmen, corporations, or individual benefactors in order to extend and focus the resources available for research. He must also maintain their trust. A successful director has to be able to persuade, suggest, and reconcile—constantly. These qualities were even more critical in the evolving institutions of the nineteenth century, especially those that depended almost completely on voluntary benevolence for support. In other words, the role of the first director of the Dudley Observatory was critical to its future.[11]

Benjamin Apthorp Gould, Jr., was completely committed to an ideology of professionalism that insisted on higher standards of training, research, and employment in order to establish a publicly acknowledged area of specialized competency. The professional aspirations of scientists like Alexander Dallas Bache, Benjamin Peirce, Joseph Henry, and Benjamin Gould were so clearly articulated in their letters to each other and so obvious in their actions that they have attracted enormous attention from scholars interested in the process of professionalization. Because Gould and the other three scientists who joined him in the quarrel over the observatory were outspoken mid-century apostles of professionalism, the conflict in Albany has been used as an illustrative example of the problems facing professionalizing scientists in a democratic society.

As Nathan Reingold observed, defining professionalization is a thankless task. Although interest in the topic has been widespread and discussion complex, scholarship in the field was recently described by Eliot Friedson as "an intellectual shambles."[12] Despite a weary consensus on the generalized attributes of a professional, the process of professionalization remains as nebulous at times as the definition of pornography in that few agree on what it is exactly, but all insist that we should know it when we see it. Whatever conclusions are drawn about the tailoring of American scientific institutions to their present style, the early history of the Dudley Observatory makes it obvious that those institutions were shaped by variables on many levels. Personality differences, local and professional rivalries, conflicts in motives or ideals, political differences—are all factors of developmental significance that seldom fit neatly into abstractions like "professionalization."

The quarrel over the Albany observatory suggests that although historical conceptualizations (as for example with models of the stages of professionalization) are valuable analytical tools, they can also be extraordinarily reductionist with regard to historical reality. The events in antebellum Albany do not invalidate the use of such constructs, but do indicate that some caution is in order in applying them across the board as explanations for the origins and development of institutions, especially in antebellum America. Mid-nineteenth-century attitudes toward science, the patrons of science, toward authority and responsibility within institutions, toward political influence, and the niceties of personal relationships within and among elites are all variables that must be considered in assessing the developmental matrix of American scientific institutions. The dynamics involved in the process of professionalization, the manner in which a scientific elite establishes its exclusive competency, are frequently more contingent and more political than conceptual discussions indicate.[13] Institutions in their early stages often developed along idiosyncratic lines easily blurred with the broad brush of sweeping generalizations.

The tumultuous confrontation over the Dudley Observatory offers an opportunity to assess a multiplicity of factors and motivations affecting the development of a particular antebellum scientific institution. Professional aspirations described in a letter from one eminent scientist to another offer one perspective on the professionalization of science in the nineteenth century. Behavior in the conflict over the Dudley Observatory offers another. Institutions demand group activity. Success in establishing any institution often depends on tangible foundations, such as mutuality of trust and willingness to compromise. Scientific institutions, like all others, are established and carried on by fallible individuals. Neither the rhetoric of professionalization nor the egalitarian rhetoric of antebellum America can adequately explain what happened in Albany in the late 1850s. Because this conflict was so public, we can follow those involved—scientists, patrons, trustees, and various partisans—down their winding paths of good intentions and through dense thickets of self-justifications to evaluate their actions as well as their ambitions. The conflict in antebellum Albany provides an opportunity to compare rhetoric with reality at a specific time. The early history of the ill-starred observatory illustrates the intricacy and complexity of interaction between social values and institutional forms.

Both of the elites involved in this conflict demonstrated a consistent intensity of positive purpose in regard to the institution. Each person believed he was acting for the highest motives in order to secure the future success of the observatory. This righteous certainty raises many questions. What were the assumptions by which each elite made its decisions in this conflict? Why did the quarrel escalate to such levels? What brought the two value systems into conflict? Other aspects of the internal structure of the American scientific community are also highlighted. What were the important questions for scientific research in the United States in the 1850s? Before research became institutionalized later in the nineteenth century, how did scientists finance their work or their increasingly expensive equipment? How were federal funds for science dispensed in the antebellum period? How did those funds affect particular institutions?

Institutional questions are also suggested. What was the role of a director of a scientific institution before the Civil War? What did patrons of a scientific institution expect in return for their support? What effect did individual personalities have on a developing scientific institution? How were scientific appointments made?

Political questions likewise abound. What were the relationships between American politicians and scientists at the state and federal levels just before the Civil War? How did the tensions of state power versus federal power in the late 1850s provide levers of influence that could be used within a scientific institution? Finally, how did different elites relate to each other at mid-century?

This bitter conflict had its roots in differing perceptions of the boundaries of authority and responsibility within a scientific institution. The controversy was nurtured in a diversity of individual motives, rivalries, and perhaps most intriguingly, a now peculiar sense of what was acceptable behavior for a "gentleman." Many of the participants on both sides lost their tempers and behaved quite irrationally during the battle. Yet each participant stood righteously on his rectitude and felt that he had suffered unpardonable wrongs at the hands of his opponents. Such stories of conflicting perceptions are always complicated, but they are understandably human in that they do not fit neatly into preconceived patterns of interpretation. The Dudley Observatory controversy involves many larger issues reflecting changing values and assumptions in the nineteenth century, but it is also complicated by smaller motives that sprang from pride, power, personal passions, prejudices, and pomposity.

In discussing the Paris Academy of Sciences, Roger Hahn compared scientific institutions to anvils on which the conflicting values of scientific and cultural elites were shaped into viable forms. The end results of the hammering seldom pleased either side, but enough benefits could be seen so that each accepted the limitations on visions produced by compromise. The anvil of the Dudley Observatory was certainly well hammered, but the vision shattered.[14]

CHAPTER 2

Albany

What can one say about Albany? "Maligning Albany is a very old game," notes the city's most gifted contemporary chronicler, William Kennedy. For Kennedy, Albany is a state of mind "as various as the American psyche itself, of which it was truly a crucible." As a novelist Kennedy believes that Albany's past provides a core sample for the American condition and the American character.[1] The sample can be a deep one. Albany prides itself on being the oldest existing town in the original thirteen colonies and the oldest chartered city in the United States. Much of the city's early prosperity must be credited to its obvious geographic advantages. From the earliest days of the Dutch fur trade, its location on the fall line of the Hudson made Albany an important commercial center. Situated strategically at the junction of the Hudson and Mohawk rivers, Albany was the logical "jumping off" point for settlers heading west. The city was an equally logical settling place for businessmen hoping to take advantage of that traffic. Prosperity fed itself. By 1850 Albany was the eighth largest city in the United States.[2]

Early nineteenth-century Albany was a bustling commercial center. The Erie, Mohawk, and Champlain canals opened west into the interior of the United States; the Hudson River flowed out into the Atlantic. Goods of all kinds followed a network of turnpikes and barge paths to Albany to be shipped on to other places. Even before the Erie Canal opened in 1825, Albany dominated trade to and from the interior of the nation. The completion of the Erie Canal quadrupled Albany's wholesale business within a few years and solidified the city's ascendancy as one of the great commercial crossroads of the continent. By 1830 Albany offered unrivaled facilities for receiving, storing, selling, and shipping goods in all directions.

In 1831 the first steam-powered railroad in New York, the Mohawk and Hudson, began operating from Albany to Schenectady. By 1850 Albany was the hub of the old New York Central Railroad system, created by consolidating eight smaller railroads. Erastus Corning, an Albany iron merchant of great wealth, prestige, and no small political power, became the first president of the New York Central. Corning established extensive repair yards for the Central on 350 acres in the West End of Albany, contributing substantially to the growth of iron foundries and other secondary suppliers to the railroad. The steady growth of rail traffic would doom the canal, but not for several decades.

As the flow of goods through Albany expanded in the second quarter of the nineteenth century, the city soon became the central wholesale market for the entire eastern seaboard. By 1840 Albany was recognized throughout the world as one of the great trading centers for raw materials. The city proudly designated itself the White Pine Center of the World, since more lumber was exchanged there than anywhere else. In the 1850s the Lumber District of Albany had forty-six firms, seven of which had gross revenues of over half a million dollars per year.[3] In addition to the lumber trade the railroad brought great shipments of livestock, giving Albany a bit of the flavor of a western cattle town. Immense herds of cattle, sheep, and hogs moved daily through the Albany streets on their way to feed lots and slaughterhouses, which covered twenty acres in the West End and supplied the entire eastern United States. Albany, which had provisioned the Army during the Revolution and the War of 1812, was challenging Chicago and Buffalo at mid-century to see which city would be the livestock center of the nation. In the 1850s Albany was the greatest depot for the wholesale trade in cattle in the country.

Lumber, cattle, hogs, sheep, barley, wheat, corn, rye, butter, lard, wool, and too many other essentials to enumerate traded on a massive scale in antebellum Albany. This abundance of raw materials attracted large-scale secondary industries, such as leather goods, paper, wood pulp, furniture, potash, flour milling, and breweries. Manufacturers were quick to see the immediate advantage of locating their factories close to a constant supply of raw materials and easy transport. The iron foundries offer an obvious example. Because ore was easily shipped by water, some of the largest foundries in the nation grew up in Albany during the second quarter of the nineteenth century, enabling the city to put up a sign proclaiming itself, among other things, as the "stove-making center of the western hemisphere." Later, of course, the foundries would leave Albany to follow the ore supplies west.[4]

This obvious concentration of manufacturing and commodity markets stimulated an equivalent expansion in banking and insurance facilities. Local demands for capital were extensive and subject to seasonal fluctuation. As grain came down the canals to Albany to be sold after the autumn harvest, local merchants and traders, including the many breweries and flour mills of Albany, needed short-term loans from banks to fund their heavy seasonal purchasing. For example, Albany was the largest wholesale market for barley in the world, but that entire trade was completed within two months each year, adding to the pressure on local financial resources. Adequate banking and insurance facilities were vital to Albany's commercial success.

In everyone's estimation the "great banker" of nineteenth-century Albany was Thomas Worth Olcott, president of the Mechanics' and Farmers' Bank from 1836 until 1880. Olcott had been the astute financier of the "Albany Regency," a political clique that dominated New York politics for decades in the early nineteenth century and put Martin Van Buren into the presidency. When the state legislature decided to regulate the banking industry in New York to ease the boom-bust cycle of the smaller banks, Olcott designed and implemented the first subtreasury or central banking system in the United States. His own Mechanics' and Farmers' Bank of Albany was named by the legislature as the profitable central clearing bank for the entire state. With capital of under half a million, Thomas Olcott exercised banking influence second only to that of Nicholas Biddle in the Bank of the United States. Obviously, if geography conferred a naturally profitable advantage on Albany, so too did capital city politics.

New York was the wealthiest state in the Union in the first half of the nineteenth century, and Albany was its capital. Like other state capitals, politics pervaded the city. The active winter social season coincided with sessions of the courts and legislature. The combination of political power with large concentrations of capital arising from local manufacturing enabled the Albany business community to expand its influence rapidly. Governors, supreme court justices, state senators, and other politicians took an active interest in local affairs, seizing the chance to make a profit in business opportunities. Members of the state legislature willingly underwrote measures that enhanced the continuing commercial prosperity of the capital city. The Erie Canal and the railroads, both heavily political enterprises, were only two obvious examples of the close relationship between politics and commerce in Albany.

Although the streets of the city were filled with cattle, hogs, and freely roaming pigs, Albany apparently retained much of the sober character and gravity of its Dutch heritage. Charles Buckingham, an English visitor who passed through in 1838, noted the contrasts between social life in Albany and New York City. There was less formal visiting in Albany, not much theater going, fewer expensive parties, and little emphasis on fine carriages or fancy dress. People in Albany, he observed with surprise, actually seemed to make a positive effort to live within their means. He attributed this to the mingling of Dutch influence with Puritan ancestry. The result, according to Buckingham, was "a more quiet and sober air to everything." That quiet and sober air in Albany society might be described as dull by less charitable observers. Boston suffered many of the same judgments when comparisons were made with the more flamboyant lifestyles of Manhattan.[5]

Albany was a good-sized, vigorous city, strong in learned institutions in the 1830s and 1840s. An active provincial culture was reflected in its museums, book stores, public lecture series, and its libraries. The Albany Academy and the Albany Institute were especially treasured by local residents. Both were founded and supported by a cultural elite that included politicians, landowners, and local businessmen. The Young Men's Association for Mutual Improvement was established in 1833 as part of the self-culture movement and set the pattern for similar organizations throughout the west. Lectures were presented twice a week during the winter season, with topics drawn from a wide spectrum, although great stress was placed on science. The association drew lecturers with substantial reputations from considerable distance and gathered its own library.[6]

Like Boston, Albany had an engrained tradition of patrician rule. Class lines were sharp in Albany. Wealth alone did not confer status. Certain families, such as the Schuylers, Livingstons, and Van Rensselaers, were accorded a deference that implied much more than mere money. These families were members of a kind of native "aristocracy," closely comparable in position to the Winthrops, Cabots, or Higginsons of Boston. But in all stable societies, the elite includes self-made men as well as aristocrats. During the early decades of the nineteenth century, new forms of economic enterprise—banking, manufacturing, insurance, and transportation—created great new fortunes. Like Boston, the men listed as directors of the banks and insurance companies of Albany were often the same individuals who served as trustees of the city's voluntary organizations, forming a kind of interlocking directorate that encompassed the important commercial, financial, and charitable institutions of the city.

Conflict in one area could lead to tensions in another. Occasional disputes over the procedures taken in the prestigious Albany Rural Cemetery Association, for example, follow the lines of competing banking groups in the city. One area of local rivalry is especially worthy of note in relation to the conflict over the observatory. It was not unusual in the nineteenth century and is still common for communities to develop business and political alignments focused upon competing commercial banks. Examples can be seen today in Chicago (First National Bank/Continental Illinois Bank), Dallas (First National Bank/Republic National Bank), Oklahoma City (First National Bank/Liberty National Bank), and San Francisco (Wells Fargo Bank/Bank of America). This rivalry was equally important in antebellum Albany. Although bank boards are not necessarily poised for immediate conflict, they are often the bases of competing centers of power. Erastus Corning and Thomas Olcott were dominant personalities in rival banks and competed in a variety of arenas: political, commercial, and social. Corning headed the Albany City Bank; Olcott was president of the Mechanics' and Farmers' Bank. Corning was a staunch Democrat with strong local support; Olcott was an emerging Republican of considerable power in the city. Local politics in Albany was a hotbed of controversy in the 1850s, compounded by bitter political feuds on both the state and municipal level, many of which were fought out within Albany city limits. These many-tiered political and commercial rivalries cut across lines of local interests and would be carried into the conflict over the observatory.

In both Boston and Albany, the successful entrepreneurs who served as directors of banks and trustees of voluntary benevolent institutions were self-assured and proud of their role in the community. Their new accumulations of wealth led to an expansion in antebellum philanthropy, due to the strong prejudice against ostentatious consumption. Like Boston, the local elite of Albany was characterized by its simple lifestyle, its strong familial traditions, and its readiness to assume civic responsibility and leadership. Men like Stephen Van Rensselaer III, known as the "Old Patroon," and Thomas Olcott, the banker, consistently demonstrated a willingness to step forward with commitments of both money and time to encourage the cultural institutions of the city.[7]

Antebellum philanthropy is difficult to trace because of the haphazard records remaining, but the magnitude of Stephen Van Rensselaer's generosity was certainly extraordinary. Few could match him in donations, but his interest and enthusiasm encouraged emulation on a lesser scale. The "Old Patroon" also demonstrates the commitment that members of the cultural elite of Albany had to supporting scientific work and establishing scientific institutions of demonstrably high quality. Stephen Van Rensselaer III was one of the first great patrons of science in the United States.

In 1820 Van Rensselaer funded one of the very first comprehensive natural history surveys in the nation, hiring Amos Eaton to survey Albany and Rensselaer counties. He then extended the survey along the entire line of the Erie Canal, paying Eaton and his assistants to give lectures in chemistry and geology as they went. Frustrated in organizing his survey by the shortage of competent engineers and surveyors, the "Old Patroon" established the Rensselaer School in Troy in 1824 and put Eaton in charge of it. Excluding West Point because it was funded by the government, the school bearing Van Rensselaer's name was the first privately established "scientific college" in the United States and awarded the first degrees in civil engineering in an English-speaking nation. Stephen Van Rensselaer's willingness to underwrite completely an institution established for professional, technical, and scientific education set a remarkable standard of sophisticated patronage for the period. Although he belonged to a dying social order, the Old Patroon embodied an aristocratic tradition that assumed that men of a certain status were preordained to be leaders in public service and cultural development.[8]

As the Old Patroon symbolized the patrician assumption of civic responsibility in Albany, business and professional men also devoted themselves to the cultural institutions in the city. A personal commitment comparable to that of Van Rensselaer can be seen in the work of Dr. James H. Armsby. From the time of his arrival in Albany in 1830, Armsby organized public support for a variety of specialized institutions. To disspell widespread prejudice against dissections, Armsby gave a series of public lectures in the 1830s, illustrated with dissections and attended by leading citizens and members of the state legislature. He aroused enough popular interest to establish a medical school in the city. With his father-in-law, Alden March, Armsby raised most of the money for the medical school and single-handedly persuaded the City Council to donate an unoccupied school building for the new institution. The new medical college was set up, but had no charter, no power to confer degrees, and no likelihood of achieving that status because of strong opposition from other medical schools in the state. Armsby solved this problem by once again involving the public and members of the legislature. A series of lectures in the anatomical theater of the new medical college generated enough support to achieve his goal. The state legislature granted a charter in 1839 and went on to appropriate twenty-one thousand dollars to the Albany Medical College, giving full credit to Dr. Armsby for his efforts and enthusiasm.

Armsby was one of that special group of enthusiasts, like Dr. Daniel Drake of Cincinnati, who were energetic institution-builders. In addition to the Medical College, Armsby was also the driving force behind the first hospital in Albany, convincing the city council to donate a building and raising over one hundred thousand dollars in donations through his efforts. In 1850 Armsby was one of the principal organizers in the effort to establish a law school in Albany and, when the charter was granted in 1851, once again the state legislature gave major credit to Dr. Armsby. When it looked as if the unendowed law school would have to suspend lectures in 1854, Armsby raised an endowment and enough extra money to build a lecture hall behind the Medical College, convincing the trustees of the medical school to donate the land for the building to the new law school. The indefatigable physician went on to draw up the organizational plans for the University of Iowa and, after the Civil War, was the major force behind the establishment of a soldiers' home in Albany.[9]

Men such as Armsby, Van Rensselaer, and the many others

who donated money and worked with them in Albany were not opposed to attempts to upgrade professional standards in the nineteenth century. By 1850 members of the Albany cultural elite had already demonstrated their willingness to organize institutions encouraging specialization and professional education. By nineteenth-century standards Albany had a remarkably successful record of philanthropic support for institutions advancing professional goals.

In addition to a strong sense of civic responsibility, the cultural elite of Albany demonstrated consistent support for scientific research. Again, as with local business undertakings, much of that interest was shared by legislators who spent time in the city. Eaton's natural history survey, funded by the Old Patroon, struck a responsive chord in the legislature. New York was not the first state to fund a geological and natural history survey, but the amount of money eventually appropriated would be truly exceptional. Although the survey had its origins in practical economic interests, the work begun in 1836 would continue under various auspices through the rest of the nineteenth century. By 1894 the legislature had appropriated over one million dollars for the survey and the publication of reports from its work. The amount expended, when compared with other states, indicates a sustained support for science over a period of considerable time within the state legislature and in Albany.[10]

The well-organized yearly reports and lavishly illustrated final reports of the New York Survey attained an outstanding reputation that lasted for generations in Europe as well as in the United States. The first volume, published in 1843, became a classic source for American geology and made the name of Albany "synonymous with science." When the great English geologist, Sir Charles Lyell, stepped off his ship in New York City on his first visit to the United States in 1845, the story was told that his first words were "Which way is Albany?"[11]

The Natural History Survey stimulated interest in science in many ways in Albany. Survey geologists often met in the homes of interested local citizens to thrash out their procedures and problems of terminology. At one of these Albany meetings, the

Association of American Geologists was founded. This organization eventually expanded to become the American Association for the Advancement of Science.[12] James Hall, a graduate of the Rensselaer school, was one of the young men who worked on the geological survey. In 1843 Hall was appointed state paleontologist and undertook the monumental task of describing and illustrating the vast collection of invertebrate fossils that had been gathered. It would take thirteen volumes and a lifetime. During his tenure as state paleontologist, Hall maintained a network of relationships with amateur and professional geologists and paleontologists throughout the nation. Hall was also a consummate politician. His position as the head of a state-funded scientific enterprise required continuing contact with influential people in the capital as well as with other professionals. Hall and Dr. Armsby worked closely on several projects, and Hall's work was warmly supported by Thomas Olcott, Amos Dean, and other members of the local elite.

Although he dealt with things long since dead, James Hall was a volatile man with a hot temper. His personality can be compared to a lightning rod in that he attracted fiery attacks and responded with bursts of explosive light and heat. As state paleontologist, Hall generated a lot of local interest in science in Albany. This interest was stimulated by the many conflicts and quarrels that periodically flowed around his controversial personality. In 1849, for example, Hall and Louis Agassiz, a Swiss scientist who emigrated in 1846 and became one of the most revered scientists of antebellum America, attacked a set of geological charts prepared by James T. Foster for use in New York schools. Foster based his charts on a description of the Taconic system written by Ebenezer Emmons, another member of the geological survey. Both Hall and Agassiz openly disparaged Emmons's descriptions and were sued for libel and damages by Foster. The sensational trial was held in Albany before a local jury. Twelve Albany citizens had to decide which descriptions of the Taconic were scientifically appropriate. The witnesses included the greatest American scientists of the day; this war of words was won by James Hall.[13]

Sensational events like the Foster trial drew increasing attention to Albany from both scientists and the general public. Albany already had a mid-nineteenth-century reputation as a center for scientific activity. Louis Agassiz spoke repeatedly of the impressive intellectual center that had arisen in the capital because of the natural history survey. During the second quarter of the nineteenth century, Albany demonstrated a consistent encouragement for professional education—as seen in the medical college and law school—and equally consistent financial support for serious scientific work. Local enthusiasm for James Hall's work drew the speculative eyes of many scientists who wished to raise the level of education and research in American science. At mid-century Albany had much to attract the particular interest of the most articulate proponents of the new ideology of professionalism. Albany, they hoped, might be the place to establish an American university for graduate education worthy of the name.

CHAPTER 3

A Scientific Elite Comes to Albany

The middle years of the nineteenth century were a transitional time in which American attitudes and institutions underwent substantial restructuring. Older social patterns had collapsed, but the new institutional forms that would provide order and authority for the future were not yet firmly established. This realignment proceeded in a somewhat piecemeal fashion in the two decades before the Civil War and was often characterized by ambitious proposals, in science as well as in the rest of society.[1]

Specialization in science was accelerating rapidly. The exploring expeditions and state surveys had encouraged specialization by presenting an immense wealth of empirical data to the scientific community and by offering scientists employment other than teaching. During the 1840s and 1850s the increasingly specialized area of knowledge known as science was being steadily fenced off from the layman, no matter how diligent a cultivator he was. Scientific associations, originally based on more democratic ideals of regional and voluntary membership, adjusted painfully. Tensions within the associations at mid-century reflected a shift in foundation, as amateurs were steadily eased out of leadership by men who thought of themselves as professionals.[2]

These self-declared professionals were attempting to define a new style in the scientific community. As a self-proclaimed elite, they wanted to establish a monopoly of competency defined by rigorous standards—often modeled on their interpretations of European science—that systematically discriminated between those who did serious and important work and those who did not. Because many antebellum scientists who evidenced the highest standards of professionalism were largely self-educated, like Joseph Henry, formal degrees meant little in distinguishing the professional from others. Nor were long lists of publications a valid means of demarcation either. Many who published extensively were amateurs who concentrated on simple descriptions of phenomena. In essence, the distinction that seems to have meant the most to the small group of scientists at the forefront of this mid-nineteenth-century movement was the evaluation they made of the seriousness of an individual's work and the rigor with which he approached it. Whether the work systematized data and extended fundamental theory made a significant difference. In consequence, those who concentrated on the diffusion of knowledge or simple description of phenomena were less esteemed. Without doubt, such judgments often carried a strong element of cliqueishness.[3]

This self-consciously professional style is most visible in the letters and actions of a small group of friends who combined recognized scientific expertise and organizational ability with control of the few truly powerful scientific positions in the nation in the 1850s. These men made up an intriguingly powerful but loosely defined group that has fascinated American historians of science as well as scholars concentrating on the process of professionalization. These scientists first designated themselves the "Florentines," then adopted the Italian word for beggars, *Lazzaroni,* in an allusion to the impecunious condition of science in the United States. Although they wrote lighthearted letters to each other, the Lazzaroni were seriously intent on raising the level of American scientific education and research. A body of literature surrounds the Lazzaroni, as individuals and as a group, for they had an enormous impact on the development of the American scientific community in the nineteenth century. Although disagreeing about the nature of their relationships with each other or the extent of their impact, most historians will concede that they shared similar aspirations for American science as individuals. Ostensibly, they were trying to change the way science was done in the United States by eliminating amateurism and by placing "the right men" in influential positions as scientific jobs came open.

Those two goals deserve some comment. Amateurism became an increasingly pejorative term in the 1840s and 1850s, especially for members of the Lazzaroni circle. Our modern sense of amateur connotes a lack of educational preparation in a particular discipline or a person with a serious avocation that does not provide his livelihood. We recognize that amateurs, from golf to astronomy, can be exceptionally competent. For the Lazzaroni, however, the word *amateur* was used to connote faulty or superficial work. The term contained an implicit judgment, not only on the work in question but also on the kind of work that scientists should be doing, as well as the style and direction of their research.[4]

The second goal, placing "the right men" in leadership positions, is much more complex in its assumptions. Many of the antebellum scientists who emphasized the diffusion rather than the advancement of knowledge were not amateurs, but they clearly fell outside the Lazzaroni definition of the right style. The same could be said of those scientists who concentrated on problems the Lazzaroni deemed trivial, such as seeking new comets or amassing collections without serious theoretical analysis. But the judgment, as made in reality, cannot be reduced to a simple question of the seriousness of a particular scientist's research. It is difficult to place the Lazzaroni wholly in the vanguard of a professionalizing movement that swept through the nineteenth century like a juggernaut because they also worked to eliminate from leadership other scientists whose work was undoubtedly of the highest professional standards—Asa Gray, the Bonds at Harvard Observatory, and William Barton Rogers. In other words, the distinctions made by the Lazzaroni in terms of where they placed their support are not equatable with a simple "professional versus amateur" dichotomy or even a "professional versus nonprofessional" judgment. In some cases the decisions that members of the circle made about who was "right" and who was not were clearly personal. In others, they reflected institutional politics, especially politics within Harvard University, as when the Cambridge Lazzaroni blocked a plan proposed by Charles Eliot and supported by Asa Gray to reorganize the Lawrence Scientific School at Harvard.

Animosities also arose over differences in methods and style in research. The Bonds at Harvard, for example, preferred the British style over the German in instrumentation and physical rather than practical astronomy. Physical astronomy concentrated on seeking knowledge of the physical properties of celestial bodies; practical astronomy emphasized measuring and computing their positions. Benjamin Gould, a close member of the Lazzaroni circle, was emphatic in his insistence on the superiority of German instrumentation, German methods, and practical astronomy. This methodological disagreement was just one of many that would affect alignments in the dispute over the Albany observatory. As a group the members of the Lazzaroni have to be included in any discussion of the professionalization of science in the United States, but their strong "us/them" judgments should not set the terms of the definition in deciding who was professional and who was not within the antebellum scientific community.[5]

Without question, the Lazzaroni revolved around Alexander Dallas Bache, the director of the United States Coast Survey, known affectionately as "the Chief." Bache has been called the "great tycoon" of American science. A great-grandson of Benjamin Franklin and nephew of the vice president of the United States, Bache was respected both in Congress and within the scientific community for his scientific work and political skills.

Joseph Henry, another important member of the circle, had grown up in Albany and was known internationally for his work on electromagnetism. "Smithson," as he was referred to by close friends, was the first secretary of the Smithsonian Institution. Both Bache and Henry have been described by Nathan Reingold as founders of the American scientific community.[6] The mathematician Benjamin Peirce, known among friends as "the Functionary," was another close comrade. Peirce was a mainstay of the Nautical Almanac and the Coast Survey, and helped establish the Harvard Observatory and Lawrence Scientific School. With a few others who shared a similar dedication—Benjamin Gould, Louis Agassiz, John Fries Frazer, James Dwight Dana, Oliver Wolcott Gibbs—they united in a concerted effort to elevate the standards and institutional foundations of American science. Members of this charmed circle wanted to participate in important scientific discussions in Europe as equals, not as provincials from the periphery. These friends traditionally gathered for evenings of good food, wine, and stimulating conversation at the yearly meeting of the American Association for the Advancement of Science. Although never in a majority on the influential Standing Committee that directed the affairs of the AAAS, they steadily gathered power through the association's formative years. Between 1848 and 1855, the association elected Joseph Henry, Bache, Louis Agassiz, Benjamin Peirce, and James Dwight Dana—all Lazzaroni—to its presidency.

The Lazzaroni goals, as Edward Lurie has pointed out, have a modern ring to them. Their avowedly professional style of science was a notable departure. Their behavior is clearly that of an elite in an emerging profession, as they worked diligently to establish internal standards of education, organization, and competency and pushed energetically for public support. Their quest for delineating marks of professional excellence reached its most exclusive and prejudiced pinnacle when this group successfully organized the National Academy of Sciences in 1863 and nominated the first fifty members. Edward Lurie has described the National Academy as the "organizational formalization of Lazzaroni values."[7] But the National Academy was still in the future when these friends convened for the 1851 meeting of the AAAS. That year the meeting was held in Albany.

Although their impressions of continental science were not always valid, American scientists complained bitterly in the 1840s and 1850s about the comparative lack of institutional support for advanced work in the United States. Because they were increasingly aware of the kind of work being done in Europe, they found their own situation unbearably frustrating. Not only were funds for research difficult to come by, but American universities (if they could be described as such) were little more than centers for undergraduate education. Advanced studies simply were not available. Not until the 1870s did Harvard and Yale eliminate the old practice of awarding master's degrees as a matter of course, provided the candidate paid five dollars and kept out of jail for three years after graduation. Even on the undergraduate level, few colleges had a teaching staff capable of giving instruction in science beyond fundamentals. If a student wished to continue his scientific education beyond this basic level, he had little choice but to travel to Europe.[8]

Visions of an American university that would rival the great German universities received enthusiastic support from both scientists and laymen in the 1850s. Richard Storr describes the mood at mid-century as a kind of "romantic academicism" in which cultural patriotism and academic aspirations merged in the idea of a university.[9] Enthusiasts insisted that an American university capable of competing as an equal in research and scholarship with the prestigious institutions of Europe would somehow focus and elevate the life of the nation. The idea of a national university in the United States was an old one that had been warmly endorsed by presidents from George Washington to John Quincy Adams. Many could see advantages in an institution that would promote higher learning and work against sectional prejudices at the same time. The many proposals had repeatedly foundered on congressional insistence that education was the business of the states, not the federal government. Despite these earlier failures, the idea of a national university emerged with new vitality in the 1850s. Benjamin Peirce even went so far as to draw up working plans for such a university. Many, like Bache, believed the future greatness of the nation depended on its commitment to advancing knowledge and increasing technological sophistication.[10]

As interest grew in the idea, cities like Albany with its history of continued support for scientific interests and institutions became a focus of attention. The New York Survey, in its organization and analysis of the field work and collections amassed between 1836 and 1845, offered an existing, funded nucleus for postgraduate instruction in many areas of science. A medical school and law school were already in operation in Albany. Many well-respected citizens were warm supporters of the AAAS, and the city was obviously in a high state of excitement at the prospect of hosting the annual meeting of the association in 1851. Albany seemed a uniquely promising place to bring the long-awaited dream of a national university into reality.

The possibility of actually creating a university emphasizing graduate and professional education was an engrossing topic among the scientists who comprised the leadership of the association. Louis Agassiz stressed his own commitment to the project in a letter to James Hall, suggesting that he persuade Thomas Olcott, the influential Albany banker, to solicit funds for the proposal from the legislature. Circulars went out from Albany announcing meetings during the winter "to discuss the subject of a National University."[11] The proposal brought the cultural elite of Albany, already noted for its enthusiastic support of science, together with a scientific elite intent upon creating institutions that would produce future generations of highly-qualified American professionals. The effort to establish a national university in Albany has been described as the first major attempt by the Lazzaroni to restructure education according to their own standards.[12]

The ever-energetic Dr. Armsby was the first to suggest that an observatory be attached to the national university proposal. The idea was received enthusiastically in Albany. As a science astronomy was thought to be the "supreme measure of the refinement of a civilization."[13] Astronomy had the added incentive of combining practical and aesthetic rewards. In addition to the obvious benefits to navigation and surveying, the contemplation of the heavens was thought to be spiritually elevating and morally uplifting. Although nineteenth-century Americans were generous in their support for astronomy, establishing an observatory was no easy task. Credit for the immediately positive response to Armsby's suggestion should undoubtedly be ascribed to the lingering effects of a dynamic lecturer who passed through the city in January of 1851.

Although he was not influential in terms of his contributions to knowledge, Ormsby Macknight Mitchel was an intriguing member of the American scientific community in the nineteenth century. His life reads more like improbable fiction than factual biography. Mitchel was an old friend of Benjamin Peirce's but was definitely not the kind of person to be invited to the self-selecting moveable feasts of the Lazzaroni. The members of the Lazzaroni fellowship were articulate proponents of a new professional style. Mitchel, a graduate of West Point and self-taught astronomer, was as well trained as his contemporaries, but he did not have the style of the future. Instead, Mitchel's career has become a scholarly stereotype for historians, one illustrating an older, preprofessional style of science that emphasized the diffusion rather than the advancement of knowledge and concentrated on the marvels of the heavens rather than on serious questions.

O. M. Mitchel was the itinerant director of the poverty-stricken Cincinnati Observatory. Mitchel had organized the Cincinnati Astronomical Society in the 1840s and built an observatory by using a unique form of donation and barter. The Cincinnati Observatory was the proud possessor of one of the finest refractors in the world, considered equal to the great telescope at the Russian Imperial Observatory at Pulkova. Unfortunately, all the money raised had been used to build and equip the observatory. No endowment was provided to support an astronomer. Saddled with debts incurred in building and running the institution, Mitchel was forced out on the public lecture circuit to support his family as well as the observatory.

Mitchel was an extraordinarily successful lecturer at a time when standards were exacting. Public speaking in the middle years of the nineteenth century was a demanding art form. Long speeches were a form of public entertainment, recognized not only as a significant art, but thought to be vitally important in a democracy as a means for providing the public with information. A great orator like Edward Everett was nationally acclaimed for his ability to establish a strongly personal relationship with his audience while informing them at the same time.[14] Mitchel excelled in this nineteenth-century art. He toured the nation each winter, giving vivid lectures on astronomy and holding audiences completely spellbound without the aid of charts or paintings. His January visit to Albany undoubtedly inspired the widespread interest in Dr. Armsby's suggestion that an observatory be added to the planned university.

Thomas Olcott wrote to Mitchel about Armsby's proposal. The astronomer suggested that they discuss the idea more fully in August when the AAAS convened its 1851 meeting in Albany. Mitchel's letter was read at a gathering at Dr. Armsby's home on 3 July at which Louis Agassiz spoke about the importance of adding an observatory to the university. Governor Washington Hunt, banker Thomas Olcott, state supreme court justices Ira Harris and Amasa Parker, and many others at the meeting committed themselves wholeheartedly to the project. In the next few days, Dr. Armsby began circulating a public subscription soliciting donations. Thomas Olcott, William H. DeWitt, and other members of the cultural elite of the city immediately pledged one thousand dollars each. In August Mitchel selected a site for the new observatory.

The sudden insertion of an observatory into the national university proposal did not please all the scientists gathered in Albany for the meeting, especially the small group who comprised the leadership of the association. Louis Agassiz supported the idea, but Bache, Peirce, and Joseph Henry were against it. They met with Olcott, Armsby, and several other supporters in an attempt to dampen local enthusiasm for the project. In their opinion money spent for another telescope that had little prospect of being used as part of a serious research program would further dissipate the already sparse financial resources available to science in the United States. Within their own respective institutions, Bache and Henry were attempting to centralize financial resources so that substantive research as well as practical work could be funded. Local endeavors like the impecunious observatory in Cincinnati or the proposed venture in Albany, as far as they were concerned, consumed funds potentially available to science without offering meaningful research capabilities in return. The United States already had two superb refractors, one at the Harvard Observatory and another at the inoperative Cincinnati Observatory. A third located in Albany, no matter how excellent, would not raise the level of American astronomical research to that of the best European observatories. With the Cincinnati equatorial as their example, they argued that another expensive instrument would simply become a nonfunctional local showpiece.

Despite the scientists' attempt to discourage the idea, enthusiasm remained strong in Albany. In September of 1851, Blandina Dudley—wealthy widow of United States Senator Charles Dudley—donated ten thousand dollars to the project in memory of her late husband's interest in astronomy, fondly recalling a honeymoon visit to the Greenwich Observatory many years before. A few days later, at the urging of Olcott and one of her nephews, she raised her initial contribution from ten to thirteen thousand dollars. General Stephen Van Rensselaer IV, sixth and last of the patroons, donated seven acres of land that Mitchel had selected as the best viewing site in the city.

In the meantime the peripatetic Mitchel, driven back on the road by his ever-pressing financial problems, brought his popular astronomy lectures back to Albany in October of 1851. At the conclusion of this lecture series, Mitchel addressed a meeting of two hundred interested citizens, speaking of both the national university and the proposed observatory. The astronomer ended his impromptu talk with a provocative challenge. If the people of Albany could not raise the money for such a grand experiment, he vowed he would do it himself. The response was electric. The meeting was brought to a fever pitch of excitement when the respected Dr. T. Romeyn Beck, former principal of the Albany Academy, rose to urge that not only an observatory but a university equal to the greatest in Europe be created in Albany. Beck's endorsement was especially persuasive. As a member of the New York Board of Regents, he had been openly doubtful about the university project before the meeting.

Mitchel's challenge that he could raise the money if the citizens of Albany wouldn't was especially provoking. Albany prided itself on its cultural institutions as well as its dominant position in trade. Cincinnati was viewed as a pushy commercial rival. Possession of one of the best telescopes in the world seemed to give a status to the younger river town that the citizens of Albany were unwilling to concede. Albany would prove its superiority by establishing a great university, complete with a renowned observatory of its own. In the following weeks, Dr. Armsby was enthusiastically received as he circulated his subscription list.

An act incorporating the national university was passed in the New York legislature. Dr. Armsby, Thomas Olcott, Robert H. Pruyn (speaker of the State Assembly), and Amos Dean of the law school were credited with pushing the university's charter through the legislature. The national university seemed closer to reality than ever before. During the fall Peirce spoke on the topic at the Young Men's Association in Albany. Other public meetings were held through 1853, but the movement did not sustain itself. The legislature refused to provide any funding. The law school and the medical school that were to be a part of the proposed university had already received appropriations, but the university as a whole did not. Despite great local enthusiasm for the idea and a steady procession of eminent scientists to Albany, the state legislature could not be persuaded to allocate any financial support.

The national university project failed for several reasons. It was obvious that the university's sponsors could not agree among themselves. Should the university emphasize teaching or original investigations, or should it attempt to do both? There was an equally fundamental division between those who wanted postgraduate and theoretical studies and those who were interested in vocational education. In addition to internal dissension, formidable opponents were assembling to defeat the idea. The New York legislature was strongly influenced by agricultural interests; the state Agricultural Society, a powerful lobbying force, wanted an institution of its own emphasizing agricultural research. The New York Board of Regents, the highest educational authority in the state, also withheld its endorsement despite Beck's enthusiasm. At the same time, the existing colleges of the state fought the proposal. Union College and Rensselaer Institute, located in nearby Schnectady and Troy, were understandably low in enthusiasm for a project creating competing loyalties, especially in fund raising. Nor were other universities, like Harvard, pleased at the prospect of losing their best scientific faculty to Albany. New York City had its own competing movement to found a university of national stature. Finally, scientists themselves were divided over the prospects of teaching for fees, as Benjamin Peirce had proposed in his organizational plan, rather than for guaranteed salaries.[15]

The problems faced by the national university were intellectual and political as well as financial. Lacking material support from the legislature, the project swiftly collapsed under these many adverse conditions. Only those existing institutions in Albany that offered recognized professional training—the law school and the medical school—survived the demise of the larger scheme. These institutions could draw on established roots of local support and commitment.

The observatory also survived. The Dudley Observatory was separately incorporated by the legislature in March of 1852, enabling it to exist as an independent entity. An initial endowment was raised and trustees were appointed. In May of 1852, General Stephen Van Rensselaer IV, the Young Patroon, was chosen as president of the board. Thomas Olcott of the Mechanics' and Farmers' Bank was named vice-president; Ormsby MacKnight Mitchel became corresponding secretary; Dr. James H. Armsby was recording secretary. Twenty-five thousand dollars had been raised. Plans for the observatory building were drawn by Mitchel, an extra acre of land was purchased, foundations were poured, and construction was completed within one year.

The observatory was located in the Arbor Hill section of Albany. Arbor Hill, its streets lined with elms, chestnuts, and fine homes, was an elegantly fashionable neighborhood, described in 1850 as one of the most delightful localities in the city. The observatory stood on the highest point in Albany, on a hill known locally as Goat Mountain because of the great numbers of the noisy animals that grazed there. Goat Mountain commanded a full view of the sky. The eight acres of the site included the plateau on which the observatory stood and an easy slope of two hundred feet down to the Hudson River. At the base of the hill, the New York Central Railroad had a line that ran up a grade. No one believed that the small engines of the infrequent trains in 1852 would present any problems to the observatory standing at the top of the hill.[16]

The observatory was a one-story brick building constructed in the form of a cross with a central dome. The east and west wings were designed to hold a transit instrument and a meridian circle. The rear wing included a library, two computing rooms, and several smaller rooms. The center hall was designed to hold the great equatorial, the showpiece instrument of astronomy. The classical design of the building was applauded for combining scientific function with architectural beauty and aesthetic effect. Everyone was quite pleased.

Everyone also assumed that O. M. Mitchel would be the director of the new observatory now standing in classical elegance on the highest ground in Albany. Mitchel certainly encouraged that impression. The trustees eagerly awaited his arrival so that he could bring the institution into working condition. Unfortunately, Mitchel's continual financial difficulties forced him to forsake astronomy completely. In September of 1852, the astronomer wrote Olcott that he was indeed leaving the Cincinnati Observatory but would not be coming to Albany after all. He was compelled to take the position of chief engineer with the Ohio and Mississippi Railroad.[17]

The Cincinnati Observatory was to be ignominiously closed on

Mitchel's departure. His predicament placed the Albany observatory in an equally vulnerable position. A sizable portion of the subscriptions still outstanding had been given solely on the understanding that Mitchel would be in charge of the institution. The astronomer could not refuse the post of director of the Dudley Observatory without releasing some of the subscribers from their commitments. Neither the Cincinnati nor the Dudley Observatory could afford to pay Mitchel a yearly salary sufficient to support himself and his family. Neither could see its future without him.

Mitchel had to remain as the nominal director of the Dudley Observatory in order to bring in the outstanding subscriptions, but both he and the trustees agreed that someone else would have to supervise the work of setting up the observatory in his absence. A young astronomer working at the Coast Survey, Benjamin Apthorp Gould, Jr., had been suggested to Mitchell as a possibility for the position, and he relayed the name to Olcott. The trustees and Mitchel conferred. Mitchel agreed to resign within a year or two after the subscriptions were paid, and Gould could then be named director of the Dudley Observatory. Understandably, after looking the offer over, Gould decided against it. The position looked insecure, to say the least. Gould was also aware that if Mitchel should return, there was little doubt where the loyalties of the trustees would lie. Rumors were circulating that Mitchel intended to keep the Albany observatory securely under his control if it appeared viable. Olcott's correspondence with Mitchel suggests that Gould's assessment was probably correct. Although Mitchel wrote Gould several times, indicating his "friendly readiness to resign" in Gould's favor as director, the younger astronomer was not persuaded. Gould expressed his gratification at the confidence shown in him but refused the invitation. Mitchel finally left Cincinnati to devote himself to the railroad. Gould remained at the Coast Survey, working on telegraphic determinations of longitude.[18]

By 1854 the Dudley Observatory still stood imposingly on its hill, a monument to Albany's willingness to support intellectual aspirations in science. Unfortunately, the building was empty.

Goats grazed freely over the hillside beyond the classic elegance of the structure. The observatory had no director and no instruments. Most of the subscriptions had been consumed in construction. The only resident of Albany's temple of science was the Irish caretaker, Joseph McGeough, who lived in the basement with his family. The Dudley Observatory had fulfilled the dour predictions made in Olcott's office in August of 1851 by Bache, Peirce, and Henry. The empty building had done nothing to advance science in the United States. Albany had done no better than its rival river city, in fact, not even as well since there were no instruments. The Dudley Observatory was as pointless an effort as the Cincinnati Observatory. Both shared the same nominal director. Both were inoperational.

During the autumn of 1854, the trustees were still trying to find a director for the empty observatory. Dr. Armsby, known for his persistence, refused to accept Gould's refusals as final. Armsby had, after all, established and "saved" both the medical and law schools. He was not about to give up on the observatory. The persevering physician continued to beg the astronomer to put the institution on its feet.[19] So matters stood when Dr. Armsby made plans to attend the ninth meeting of the American Association for the Advancement of Science convening in Providence, Rhode Island, in the summer of 1855. Armsby was obviously not a man who could be put off easily, no matter how strong the resistance. Sometime during the meeting, Armsby joined Bache, Peirce, and Gould in a private discussion about the languishing observatory and "the good which might be done" for science if they would interest themselves in it. Armsby insisted that the trustees were willing to do *anything* to see that the institution was successful. Bache and Peirce made a counterproposal of their own.

Each spoke to the physician of the need for an esoteric and inordinately expensive instrument, an astronomical measuring device called a heliometer. Armsby assured the scientists that if they would support the institution the instrument would be secured for the Dudley Observatory. In return for this guarantee, Bache agreed to make the observatory in Albany an essential

American point of longitude by using it as the primary location in the Coast Survey's triangulation of the Hudson River. Bache implied that this designation would give the Dudley Observatory the stature of an American Greenwich; it would be the final point of reference for all American meridians. In addition Bache proposed that the Coast Survey team involved in longitude work should supervise the remaining construction and installation of the instruments to be purchased for the observatory. A transit instrument and the services of an observer for the heliometer would also be provided by the Coast Survey. Before he left Providence, Dr. Armsby had secured an informal arrangement with the superintendent of the Coast Survey that seemed to promise greatness for the Dudley Observatory. Once again, an adamant Armsby had rescued an endangered Albany institution from extinction.[20]

The question that now has to be asked is obvious. Why did these scientists change their minds and decide to interest themselves in the Albany observatory? They had been adamantly opposed to the entire proposal four years earlier. Two considerations must be taken into account: first, the method by which astronomical research was funded in the United States at the time, and second, the questions astronomers wished to answer in 1855.

With regard to funding scientific work in the United States, A. Hunter Dupree has observed that every generation since 1830 has had a predominant scientific agency that appeared invincible during the period of its ascendancy.[21] In 1855 the dominant agency was the Coast Survey. The survey attained its position by managing to satisfy two important constituencies. It offered obvious practical benefits in providing precise charts for the shoreline, patterns of tides in the Gulf of Mexico, studies of the Gulf Stream, as well as a scientific approach to the standardization of weights and measures—all services of immediate value to commercial interests in the United States. At the same time, the Coast Survey systematized research on questions of basic scientific interest, such as magnetism. Because of the comparative scale of its efforts and the size of its budget, Bache's agency had an enormous impact on the American scientific community, especially in the physical sciences. The Coast Survey was particularly influential in American astronomy in that it offered instruments to existing observatories in return for accurate observations. At a time when there were no graduate schools and few facilities for scientific research, the Coast Survey was a pioneer in training scientific personnel, a continual employer of mathematicians, astronomers, and physicists. It was, as Benjamin Silliman termed it, "a school for special training."[22] Under Bache's leadership scientists throughout the nation worked for the Coast Survey, which was not surpassed as an employer of scientists in the United States until the 1880s. The Coast Survey's work was highly respected in Europe, with great interest taken there in the new method devised for determining longitude by using the telegraph, known as "the American system."

Bache always managed to see that the practical side of work was done to avoid possible criticism, but he also guided the agenda toward important scientific questions. A man as politically astute as Bache did not hesitate to use the patronage under his control to direct American science toward the European models of excellence he admired. Scientists who did not measure up to his standards found themselves without funding or employment.

Although Congress had directed the Coast Survey to determine longitude, it had expressly forbidden any of the allocated funds to be used in constructing or maintaining a permanent astronomical observatory. Support for science was thought to be the proper function of the state governments. The federally funded Coast Survey was therefore restricted to using portable astronomical instruments. Without access to a permanently established observatory, basic research on the most important questions in nineteenth-century astronomy was impossible.

At the middle of the nineteenth century, research in astronomy concentrated on the determination of some basic fundamentals of astronomical measurement. The question of stellar parallax had been important since the time of Copernicus. As stellar astronomy came to the fore with the work of William Herschel (1738–1822), new generations of telescopes and micrometers were developed so that the proper motions of stars—first discovered by Edmond Halley in 1718—and separations of double stars might be studied in an attempt to solve the parallax problem. But the difficulties of this particular investigation were, as the Astronomer Royal, George Biddle Airy, termed them, "almost bewildering. So long as we conceived any one material body to be fixed, our notions of motion, whether of translation or of rotation or revolution, are perfectly clear; but when once we give up the idea of material immobility, we soon perceived that the notion of a zero, either of linear measure or of angle, is a purely metaphysical conception. In applying it to nature, we must assume something as arbitrary."[23]

How could stellar distances be determined without a stable point of reference? Galileo had suggested that observing two stars situated close together in the sky might offer a solution. By making observations at times when the earth was in opposite positions in its orbit around the sun, the motion of one star with respect to the other could be ascertained. This was known as the double-star or differential method of parallax measurement and was used by Bessel in making the first successful determination of stellar parallax in 1838. Bessel chose *61 Cygni* for his observations because that star moved across the sky at the astronomically remarkable rate of 5.2 seconds of arc per year. Bessel measured this infinitesimal movement with an instrument created by the renowned Fraunhofer—a heliometer. Although the Königsburg heliometer had been originally designed to ascertain the diameter of the sun, Bessel saw at once that he could use the instrument to attack one of the oldest problems confronting astronomers—the distance of the fixed stars as evidenced by their annual motions. He succeeded.

The heliometer was a measuring instrument *par excellence.* The Oxford heliometer, with a seven-and-one-half-inch object glass made by Merz and mounted by Repsold in 1848, is still considered to be one of the finest precision instruments ever made. It was at the cutting edge of optical technology at the middle of the nineteenth century, the most advanced state of the instrument-maker's art. Only a few existed. Technologically sophisticated, a heliometer was a very special kind of telescope. As George Bond of the Harvard Observatory described it to Asa Gray, the heliometer was "the most complex of astronomical instruments."[24] Although it was far more accurate in measurement than a transit or meridian circle, the operational procedures were painstakingly laborious. The objective lens of the heliometer was divided so that two visual images could be adjusted with a finely calibrated set of screws whose revolutions were precisely set to fractions of a second of arc. Each half of the lens produced the image of its respective star. By moving the halves, one visual image could be brought into relation with the other. The number of revolutions required to bring the images together yielded an immediate measurement of the angular difference between the two stars. The angle to be measured was minuscule, usually less than a tenth of a second of arc, equal to that subtended by the head of a pin at a distance of two miles.[25]

The most exciting research in astronomy of the 1850s demanded instrumentation of this meticulous exactitude. American astronomers were envious of the few European astronomers who possessed instruments capable of such elegantly precise measurements. The longitude work of the Coast Survey required fastidious accuracy and was well-respected, but the context of their longitude determinations was always a derivative one, relying on someone else to establish the values of stellar and solar parallax that would fix the earth's actual position in space. The semidiameter of the earth's orbit around the sun—still uncertain at this time—was an essential base line for the trigonometric determinations of all celestial distances and a basic reference point to determinations on earth. Inaccuracy in this basic measurement made all longitude determinations tentative in an ultimate sense and robbed the Coast Survey astronomers of the international stature they were seeking.

American scientists were often infuriated by the patronizing treatment they received from their European peers. Many American scientists of this period, especially those of the Coast Survey, wanted to show their European counterparts that they could not be dismissed or discounted. Cultural nationalism was an especially strong motivating force, as seen in the movement in the 1840s to establish an American prime meridian. Many of the scientific expeditions had been patriotic in motivation. The Gilliss Expedition to Chili, for example, attempted to establish values for solar parallax and was successfully sold to Congress on the grounds that the data would be "wholly American."[26]

Not all American astronomers believed that the question of solar parallax was as pressing as the Coast Survey scientists insisted it was. Both William Cranch Bond and his son, George Bond, of the Harvard Observatory thought that the emphasis on the solar parallax question was merely a means of encouraging popular support for an increase in the Coast Survey budget so that those astronomers could work in areas in which they were interested. The solar parallax question, especially as it related to the Gilliss Expedition, provided an area of contention that festered between the Bonds of Harvard Observatory and the Lazzaroni fellowship. Benjamin Peirce strongly disagreed with the Bonds and spent much of his time on the problem of solar parallax.[27]

The sluggish proper motion of the Pleiades was another subject of immense interest in mid-century astronomy. In the bewildering interrelationships of celestial revolutions, astronomers searched for the center of galactic rotation. Although this question would not be resolved until the 1920s when Harlow Shapley discerned the center was in the direction of the constellation Sagittarius, many astronomers of the 1850s thought the Pleiades might be the center of the galaxy. Motions were so small in the Pleiades, however, that the only acceptable observations had to come from instruments of unquestioned accuracy. Coast Survey astronomers were necessarily barred from these scientific debates because of their inability to bring precise measurements to the discussion.

Without accurate measurements American astronomers knew they would remain on the periphery, excluded from the most important astronomical discussions of the time: stellar parallax, solar parallax, the center of galactic rotation. Unless Americans came supplied with observations of unimpeachable accuracy, they could not hope to participate. About this time a European astronomer, recently emigrated, was employed by the Coast Survey to work on Pleiades observations under Peirce's supervision. Dr. C. H. F. Peters had spoken to Peirce at length of the importance of a heliometer if observations of the desired precision were to be obtained. But the heliometer was a fixed astronomical instrument, not portable in any sense, and could not therefore be purchased by the Coast Survey under the restrictions imposed by Congress.

And so we come to the crux of the sudden surge of interest shown by these scientists in the lethargic career of the Dudley Observatory of Albany. The scientists at the Coast Survey wanted the measuring capabilities of the heliometer for their observations. Because an act of Congress expressly prohibited the survey from establishing an observatory, Bache proposed that the citizens of Albany purchase a heliometer. In return, "he, as Superintendent of the Coast Survey, and in consideration of the advantage he should derive from the use of that instrument, would supply a Transit instrument and observers, from his corps of United States employees, free of expense to the institution."[28]

The bargain was sealed with enthusiasm on both sides. Dr. Armsby returned to Albany with heady visions of international renown for the Dudley Observatory. His enthusiasm carried him and Thomas Olcott out to visit Mrs. Dudley. After listening to the excited discussion, she was equally ardent in her support. Generously, the widow offered to supply the full purchase price of a heliometer. This news was immediately relayed to Peirce and Bache. The two scientists promptly suggested that an advisory council be appointed for the observatory, offering their own names along with Captain Davis of the Nautical Almanac and Benjamin Apthorp Gould, Jr.

In the meantime Benjamin Gould was on his way home from the Providence meeting. When he arrived in Cambridge, he found letters and telegrams from Dr. Armsby, begging him to come to Albany as quickly as possible. Joseph Henry was visiting

Cambridge also and agreed to accompany Gould. They arrived on 3 September, were met by Armsby at the train, and hurried along to a meeting already in progress at the doctor's home. Both Gould and Henry made short speeches to those assembled, each emphasizing the pressing need for an American scientific institution respected internationally for the quality of its research. Gould discussed the advantages offered by instruments such as the heliometer and meridian circle, stressing that "accurate, careful measurements rather than . . . the contemplation of physical peculiarities" should be the focus for a superior observatory.[29]

As it was eventually settled, the prestigious Scientific Council of the Dudley Observatory included Alexander Dallas Bache, superintendent of the Coast Survey; Joseph Henry, former Albany resident and secretary of the Smithsonian Institution; Benjamin Peirce, Perkins Professor of Mathematics and Astronomy at Harvard University; and Benjamin Apthorp Gould, Jr. These men were truly a scientific elite at mid-century. The observatory's board of trustees included acknowledged leaders of a community that had consistently demonstrated its enthusiasm for science. Those trustees were strongly committed to the success of the institution. The convergence of these two elites promised greatness to the project in which they were united—the Dudley Observatory of Albany.

CHAPTER 4

Benjamin Apthorp Gould, Jr.

Three of the four members named to the Scientific Council of the Dudley Observatory possess names familiar to any who have studied the development of American scientific institutions in the nineteenth century. Alexander Dallas Bache, Joseph Henry, and Benjamin Peirce were men of massive contemporary reputation, a threesome who exerted an extraordinary influence within the scientific institutions and associations of antebellum America. The fourth member of the Scientific Council lacked comparable reputation and influence at midcentury. His name is less familiar, but the early history of the Dudley Observatory is centered, in a most personal way, around Benjamin Apthorp Gould, Jr. His ambitions—for himself and for American science—and his convoluted personality provide a pivotal point in the story.

Charles Rosenberg has described the years before the Civil War as marked by a mood of intense, even millenial enthusiasm among scientists.[1] If "professionalization" has become an organizing theme for many discussions of this period, Benjamin

Apthorp Gould certainly deserves a biography of his own. He was deeply committed throughout his life to the task of raising the level of science in the United States to European standards. Unfortunately, one of the most serious difficulties confronting any scholar who wishes to assess Gould's place in the development of nineteenth-century American science is the obvious absence of a collection of his own letters and papers. Gould was a prolific correspondent, but his own extensive files were lost, leaving little private information behind. Many of his letters, however, sometimes two a day to the same person, are scattered through the collections of his friends as incoming mail. Anyone wishing to become an expert on Benjamin Gould must be prepared to visit a wide variety of collections and, like a hunter for hidden treasure, will constantly unearth new letters in unsuspected places. Given the complexity of Gould's personality, this almost seems appropriate. Nothing about Benjamin Gould was ever easy.[2]

Benjamin Apthorp Gould, Jr., was born in Boston on 27 September 1824, the eldest of four children. Gould had a lifelong interest in genealogy, especially his own, priding himself on his descent from Puritans arriving in the 1630s on both sides of his family. He even compiled an extensive family genealogy, which was printed twice.[3] His father, Benjamin Apthorp Gould, Sr., served as headmaster of the Boston Latin School from 1814 until 1828 when his failing health forced him to leave. The senior Gould decided to recuperate in Europe, but returned to Boston later as a merchant and shipowner, trading with China and the East Indies.

During his parents' long absence, Gould remained in Boston with his aunt Hannah, who was known locally for her poetry. Benjamin Gould, Jr., was a precocious child, reading aloud by the age of three, composing odes in Latin at five years, and giving lectures on electricity, illustrated with machines of his own construction by the time he was ten years old. After attending various primary schools, Gould entered the Boston Latin School. An excellent student, he went on to Harvard College at age sixteen, rooming with Francis Parkman, the future historian, in Holworthy on Harvard Yard.

Science and mathematics soon overshadowed Gould's earlier interest in the classics. Although he was at first interested in biology, Gould devoted himself increasingly to astronomy as he studied under the supervision of Benjamin Peirce. When a friend remarked on his unexpected knowledge of trees much later in life, Gould replied that he had narrowly escaped being a botanist. During Gould's undergraduate years, the Harvard Observatory joined in a project proposed by Baron Alexander von Humboldt, an international effort to compare simultaneous observations on magnetism by German, Russian, and British scientists. The American Philosophical Society petitioned Congress to establish five magnetic observatories in the United States as part of the magnetic crusade, but the proposal was rejected in 1839. Despite this lack of public support, individual institutions like the Depot of Charts and Instruments (later the Naval Observatory) and the Harvard College Observatory decided to participate anyway. Magnetic observations were made for three years at Harvard, from 1840 through 1843. The Harvard Observatory relied on volunteers for the project, a group of students who called themselves "The Meteorological Society of Harvard." No records attest to Gould's direct involvement in this project, but he did spend free time at the observatory as an undergraduate while these observations were being collected.[4]

Family business difficulties forced Gould to interrupt his education between his junior and senior years. During this hiatus Gould taught briefly in Lexington, Massachusetts, and at the Roxbury Latin School, then returned to Cambridge to finish his degree. At the age of nineteen, he graduated in 1844 from Harvard with a distinction in mathematics and physics as well as membership in Phi Beta Kappa. After graduation Gould returned to the Roxbury Latin School. Within a year he knew he had no desire to continue teaching. Seeking advice from Sears Cook Walker, a family friend with an interest in astronomy, Gould began to prepare for a career in science. Walker suggested that if he were serious about devoting himself to scientific study, Gould should spend time in Europe, emphasizing that only by learning European languages and methods could he keep pace with science.

For Americans interested in science in the 1840s, the European universities beckoned irresistibly. In explaining why Americans were so willing to pursue advanced studies abroad during this time, Charles Rosenberg suggests that some wished to pursue a nonmaterialistic career with a higher purpose. Others had an unquestioning faith in progress or a nationalistic desire to bring the best back to the United States.[5] Gould's decision reflects all of these motives. Leaving behind his teaching salary of $1,250, he sailed for Europe in July of 1845. His decision was an uncommon one. As he described it himself, "I believe I may say that a single instance of a man's devoting himself to science as the only earthly aim and object of his life, while unassured of a professor's chair or some analogous appointment upon which he might depend for subsistence, was wholly unknown."[6]

Family connections furnished him with letters of introduction from President John Quincy Adams, Benjamin Peirce, and Augustus Addison Gould that would open doors for him all over Europe. The trip would be the great watershed in his life. Gould spent four years in European observatories, meeting the most eminent scholars of Europe—George Biddle Airy, François Arago, Friedrich Wilhelm Argelander, James Chaliss, Carl Friedrich Gauss, H. C. Schumacher, Alexander von Humboldt—and establishing a network of future correspondents which would last as long as he lived.

Astronomers of the nineteenth century learned by doing. If Gould intended to become a serious astronomer, he knew he had to acquire practical experience in observatories. Gould first introduced himself to George Biddle Airy, the Astronomer Royal of England, and spent three months at the Royal Observatory at Greenwich. Airy was described by Simon Newcomb as the most commanding figure of his time in astronomy. As Astronomer Royal, Airy was respected both for his mathematical contributions to astronomy and for his administrative methods at the Greenwich Observatory. Airy's functionally organized innovations reflected the prevailing values of an industrial society—order, discipline, and efficiency. In effect, Airy introduced large-scale production into astronomy. The Greenwich Observatory was set up on the lines of a mid-nineteenth-century astronomical factory. Instruments were massive, and the division of labor was precisely organized. In a functional sense, an astronomer under Airy's rule could be compared to a factory hand. He was assigned a closely circumscribed task and given a specific procedure to follow. Some objected to the regimentation, but no one could argue with the fact that Airy made the Greenwich Observatory one of the most productive research institutions of the nineteenth century. Airy's emphasis on discipline and order flowed naturally from his autocratic personality. As Astronomer Royal, Airy emphasized the importance of the observatory's contributions to commerce, making his dislike of theoretical investigations obvious by continually criticizing other astronomers for the impracticality of their work. His style did not appeal to Gould.[7]

From Greenwich, Gould moved on to the Paris Observatory to work for four months with François Arago and Jean Baptiste Biôt. Although the work was interesting, he had not yet found his intellectual home. Joseph Henry had observed almost a decade earlier that the French scientific community was preoccupied with politics and economic considerations rather than with research.[8] The French style did not suit the young American, so he set off to Berlin where Baron von Humboldt and the American envoy in Berlin persuaded J. F. Encke to take Gould on as a student at the Berlin Observatory. He remained for a year.

Germany captured Gould's imagination completely. In contrast with staid English attitudes toward education, German universities experienced a kind of revolution in the early nineteenth century. Wilhelm von Humboldt, brother of Baron Alexander von Humboldt, implemented reforms that restructured the entire German educational system. These reforms combined with the founding of the university of Berlin in 1810 to create a new institutional environment for scholarship. Inspired by German idealism, professors competed with each other in a Faustian search for pure truth and new knowledge. Concentration on research emerged first in philology seminars, then in history, spreading out to other disciplines. The universities of Halle and

Göttingen gained an early preeminence, but Justus von Liebig's laboratory at Giessen set the pattern for research in the natural sciences. The Prussian universities of Berlin, Breslau, and Bonn acquired equally impressive reputations. By 1850 the idea that the university should be, above all, a workshop of free research prevailed in Germany, drastically accelerating specialization, especially in science. The university of Berlin was a striking representative of this German ideal.[9]

Completely enraptured, Gould described Berlin as a "great scientific emporium . . . whence scientific correspondence and means of communication are going on continually with all parts of the continent." He relished the atmosphere of the German scientific community, which suited his temperament completely. As he phrased it, he preferred to be where "scientific men fight like cats and dogs" and gleefully contrasted the pugnacious attitudes of German scientists with English reticence to attack the scientific work of others for fear of retaliation.[10]

Gould carried this preference for the German style of science over the English for the rest of his life, criticizing British influences on American science in general and astronomy in particular. As a young man, he made a lifelong commitment to German methods in astronomy, methods that emphasized a rigorous mathematical approach to both problems and instruments.[11] In addition to a set of techniques and concepts, the German experience gave Gould an attitude toward his discipline that set the terms of his scientific life. The German professor who dedicated himself to research and disdained material rewards was a formative image. For Gould, Germany would always be the true home of the exact sciences, the "center" of the scientific world. German standards became Gould's frame of reference when he thought about the proper conditions for scientific work. Germany established the criteria by which he would judge a scientific community in the future.[12] During this time in Berlin, Gould also forged a lifelong friendship with Oliver Wolcott Gibbs, a young American equally enthralled with the style and standards of science in Germany.

While Gould was working as an assistant to Encke at the Berlin

Observatory, Professor Galle made the dazzling visual discovery of Neptune, using that observatory's telescopes to follow Leverrier's calculations for the elusive planet's presumed position. Gould kept notes and would write his own narrative of the discovery when he returned to the United States.[13] During this exciting year, he cemented long-lasting friendships with two eminent European scientists—Baron Alexander von Humboldt and Friedrich Argelander. Through Humboldt's influence Gould was given an opportunity beyond his best hopes. He would continue his studies at the University of Göttingen under the supervision of Carl Friederich Gauss.

Gauss is frequently described in the twentieth century as the founder of modern mathematics. At the middle of the nineteenth century, Gauss was generally acknowledged to be the greatest mathematician since Isaac Newton. In addition to his stature in mathematics, his contributions to theoretical and experimental magnetism as well as astronomy were themselves enough to reconstruct these sciences along new lines of research. The universality of Gauss's activities was especially remarkable since his German contemporaries were becoming increasingly specialized.

In March of 1847 Gould wrote Gauss to ask if he could come to Göttingen to study under the mathematician's direction, forwarding a letter of recommendation from Humboldt. By April Gauss had agreed. The opportunity was a rare one, unprecedented for an American. Gauss not only took Gould on as a student but also welcomed him into the rarified atmosphere of his inner circle of friends. Gould described Göttingen as "ten times better" than Berlin. He remained for one year and received his doctorate in 1848.[14] Gould left Göttingen and made a leisurely journey home, visiting important European observatories along the way. He spent four months at Altona with Schumacher, the respected editor of the *Astronomische Nachrichten* and one month at Gotha Observatory. Moving through political revolutions in Russia, Bohemia, Hungary, and Italy, he returned to the United States in December of 1848.

The young astronomer had experienced three years of intensive immersion in the heady academic traditions of German universities. During his time in Germany, Gould internalized a style of "doing science" that became a significant part of his own self-image. He returned to the United States with firm attitudes—about the nature, methods, and purposes of proper astronomical research—constructed on the foundations of this intense German experience. When Gould called himself a "professional astronomer," he thought in European, and even more significantly, in German terms.

The contrast with American conditions was stark. In a letter to Augustus Addison Gould in 1847, Gould complained that Americans talked a great deal about perseverance under difficulties but didn't have the slightest understanding of the enormous obstacles that confronted "the scientific man" in the United States. As Gould later described the American situation in the 1840s, many circumstances worked against the development of an active scientific community in the United States. The dearth of proper libraries, lack of interest in foreign languages, and unfamiliarity with the science of continental Europe "except as presented in the garb of British translations and commentaries" tended to delay the development of an intellectual independence commensurate with the political independence enjoyed by Americans.[15]

To Humboldt, Gould confessed himself stunned by the full extent of the problems confronting American scientists. He felt compelled to do something to improve the state of science in the United States and personally dedicated himself to raising the reputation of American astronomy. In the tradition of his Puritan ancestors, Gould acknowledged his calling, the unshirkable duty that lay before him. He wrote Humboldt that he was dedicating all his effort, "not to the attainment of any reputation for myself, but to serving to the utmost of my ability the science of my country—or, rather as my friend Mr. Agassiz tells me I must say, science in my country."[16] He wrote in a similar vein to Encke in 1849, saying that his only aim was to show through perseverance and determination that he placed "a higher vaslue on the true improvement of our American science than on personal comfort, salary or reputation."[17] It was not easy. Although astronomy seems to be an "ivory-tower" science, economic, political, ideological factors, and national aspirations have always shaped the lives of astronomers and influenced their productivity. There weren't many positions open for a young man with a doctorate in astronomy in the late 1840s, even if his degree came from a prestigious university like Göttingen.

When Gould left Europe, he was fluent in French, German, Spanish, and his Italian was passable. He supported himself for two years in Cambridge by giving lessons in French, German, and mathematics. Although the contrast in scientific environment and conditions between Germany and Massachusetts required a difficult adjustment, he was not discouraged. He continued his efforts to establish himself as a scientist by European standards. Between 1848 and 1851, Gould published more than twenty scholarly articles on the motions of comets and asteroids. Even more significantly, Gould used his own funds to launch a professional journal for astronomers in 1849. His motives stemmed from his aspirations for American science. Keenly aware of the function of a scientific journal in stimulating research, Gould had noted the dearth of American publications in European libraries during his stay in Europe. Others, like Joseph Henry, had earlier lamented the absence of a truly scholarly publication in the United States.[18]

Without a journal providing a means for communicating research over a wide geographic area, American scientists were inevitably out of touch with recent developments in their fields. Astronomy, more than any other science, required long-distance cooperation and frequent comparisons of observations and computations. On the American side of the Atlantic, O. M. Mitchel's *Sidereal Messenger* was the only astronomical journal available. Mitchel's publication was completely unsatisfactory for professional astronomers, as Mitchel directed his articles toward the interests of amateur enthusiasts.[19]

Gould patterned his *Astronomical Journal* on the most prestigious scholarly journal in Europe, Schumacher's *Astronomische Nachrichten,* even copying the form and layout of Schumacher's publication. In pointed contrast with Mitchel's *Sidereal Messenger,* Gould took the highest ground in his own editorial policy. No works of a popular nature would ever be included. Articles were to be solicited from professional astronomers, and Gould was adamant that only "original researches" would be published. Nor would any translations of European articles or reprints of previously published information be accepted. Gould's *Astronomical Journal* was to have "nothing for the diffusion,—as much as possible for the advancement of science." In the preamble to the first issue, Gould asserted that the *Astronomical Journal* was absolutely necessary for the "proper development of astronomy" in the United States.[20]

Although the *Astronomical Journal* was highly praised by European astronomers for the level of its scientific publications, it was slow in appearing and extremely expensive. Astronomers needed to compare observations frequently. They also operated on slim budgets in the United States. Some American astronomers, such as George Bond of the Harvard Observatory, were of the opinion that Gould was less than fair in deciding who would be published and who would not. When Dr. Friedrich Brünnow later began publishing his *Notices* as an alternative, Bond and other "non-Lazzaroni" astronomers supported the publication in hopes that it would be more serviceable as well as accessible. But Brünnow's *Notices* were not published regularly, and Gould's publication continued to be important.[21]

The *Astronomical Journal,* excellent though it was, remained a constant drain on Gould's finances. Printing costs ran about $900 per issue and the total income was only about $1,025. By 1853 he was "very gloomy about the Journal" and spoke of discontinuing it. He wrote to the Astronomer Royal in 1854, complaining that the "support of such an expensive publication is no small load for a young man."[22] Although Benjamin Peirce suggested that Bache channel government publications on astronomy through the *Astronomical Journal,* Gould's financial condition had not improved by 1855. The discouraged astronomer felt that he was giving most of his time and energy to keeping himself and his publication in existence.[23]

The *Astronomical Journal* provides an almost symbolic example of the strength of Benjamin Gould's devotion to science. His editorial policy reflects the totality of his commitment to his vision of scientific excellence. George Daniels singled out the young Gould as one of the most striking examples of the new ideal of science-for-its-own-sake in the mid-century scientific community. Bache, Peirce, Agassiz, and Joseph Henry and other scientists of the antebellum period certainly exemplified similar values. Yet, because of the intensity of his dedication and the absence of any willingness to compromise, the impression Gould leaves is more starkly drawn, unrelieved by shadings or softened edges. Because his sense of his own professionalism was untempered by the astute political abilities of someone like Bache, Gould lends himself to stereotypical interpretations of professionalism, as O. M. Mitchel does on the opposite end of the continuum. Gould's values are more acceptable to modern observers than Mitchel's, for they foreshadowed the institutional style and forms of twentieth-century science.[24]

Benjamin Gould should not, however, be viewed as a twentieth-century scientist in the wrong place and time. He was very much a nineteenth-century person, both in attitudes and behavior. In the nineteenth century, without a supporting institutional framework, this new professional style demanded an extraordinary intensity and dedication. Benjamin Gould's personality was truly centered in the totality of his commitment to science but in a nineteenth-century context. Instead of thinking of Benjamin Gould as a man ahead of his time, the image of the Christian missionary of the nineteenth century is perhaps a more appropriate analogy. Benjamin Gould might best be characterized as a kind of missionary-for-science, closely comparable in intensity to those intrepid idealists who carried the message to China and Africa during this same period. Like those stouthearted and often single-minded Christians, his entire professional career has a coherence of intellectual definition that can only be described as teleological.

During 1851, as he labored to support himself and the *Astronomical Journal,* Gould received an unprecedented offer. He could take up a professorship at Göttingen University, with the sworn promise from Gauss that he would become director of the Göttingen Observatory. If he accepted Gould would be the first American to become a professor at a continental university as well as the first American to become the director of a European observatory of significant reputation. With such a position, his life could be productive in the fullest scholarly and scientific sense. Gould's friends urged him to accept. He did. Then, a few days later, he wrote another letter refusing the offer.[25]

At this point it is clear that Gould committed himself totally to his own mission—to raise the level of science in the United States to European, or more particularly German, standards. Gould's identity and his perception of himself were enveloped in his sense of this personal mission. Though at first tempted, the choice was not difficult for a man who prided himself on his ancestor's willingness to risk everything to create a more perfect society in the new world. With his unique educational background and intense dedication, Gould was convinced that he could make a substantive difference in his own time. He intended to provide the special leadership needed to raise American standards of scientific work to European levels of excellence.

Gould had been a protégé of Benjamin Peirce while an undergraduate at Harvard College. Through Peirce, Gould became a loyal member of that small group of scientists who shared his goals for the future and called themselves the Lazzaroni. The Lazzaroni worked hard, if occasionally tactlessly, in their efforts to reshape the diffuse membership of the American Association for the Advancement of Science into a more professional scientific society. Although often discouraged with that unwieldy organization, Gould and his friends hesitated to abandon the AAAS completely for fear that something worse would take its place.[26]

In the meantime Bache gave Gould a position in the Coast Survey that enabled him to support himself and his journal. He was to take charge of the telegraphic determination of longitude, a project begun by Sears Cook Walker, who was then terminally ill. From 1852 until 1867, Gould described himself as the head of the longitudinal department of the Coast Survey. He had complete charge of all the stations engaged in longitudinal determinations. Later, in a dispute with his mentor Peirce over leadership of the Coast Survey, Gould described his feelings about that time in his life:

> To this work I gave the almost exclusive labor of the best years of my life, working unremittingly and trusting to the future for whatever recompense might be awarded me, when the results should be published.
>
> In all this time, no question of official relation, position, or rank ever arose, nor did it ever occur to me that such questions could ever arise. The fact that the telegraph matters were solely in my charge was patent, only for such an end was I connected with the survey, my profession being that of astronomer, not of an engineer, and if I never claimed an official title or official forms, it was because such matters are indifferent to me.[27]

The longitude work under Gould's supervision provided him with government funds for working out basic questions in electromagnetism—the velocity of signal transmission and measurement of the personal equation involved in telegraphic signals.[28] Much of Gould's work for the Coast Survey was in the field, and his letters to his friends—Peirce, Bache, Wolcott Gibbs, John Fries Frazer—are filled with stories of the hardships endured.

In discussing the process of professionalization, Burton Bledstein has noted that the middle-class personalities involved were characterized by a commitment to handle life, to overcome it, to "stick it out." In addition to toughness, Bledstein portrays the middle-class professional of the nineteenth century as a person who needed to be publicly recognized as someone with a special gift and a decisive influence over others. Bledstein describes relentlessly competitive men to whom it was important to go it alone and "skunk" the opposition.[29] Gould's letters are almost stereotypical for these descriptions of the personality type of the antebellum professional. His notes to his friends are filled with military analogies and stories of adversities overcome through willpower and determination. But Gould's assertions of competency and toughness had a soft underside of self-doubt, causing periodic lapses into inertia, confusion, indecision, and depression.

During 1851 Gould was clearly despondent. He announced that his poor health made it necessary to turn the editing of the *Astronomical Journal* over to Benjamin Peirce and Joseph Hubbard for a while. The following year he again complained frequently about his ill health and the pressure of work that prevented him from ever getting a good night's rest. By spring of 1854 Gould wrote Airy that his worries over the *Astronomical Journal* and the pressure of his work at the Coast Survey had depressed him terribly, "owing in part, to my own exhaustion and not very hopeful state of mind." Gould made repeated references throughout his life to incapacitating cerebral attacks that prostrated him. During the 1850s his precarious emotional condition was described as fragile by his friends. Bache seemed to think the causes were "mental."[30]

At this point some conjecture about Gould's mental condition is warranted by the course of coming events. Although many hours were spent discussing Benjamin Gould with a clinical psychiatrist who was exceptionally generous with his time, this speculation makes no pretensions to any first-hand form of psychohistorical expertise.[31] Any attempt to offer a definitive psychoanalysis of someone who not only has been dead for close to a century but whose own personal records have been lost and/or destroyed is doomed to be buried under a weight of justified criticism. Nevertheless, Benjamin Gould's behavior and his state of mind deserve some comment, for they are pivotal to the conflict that followed.

The arrogance that characterized much of Benjamin Gould's behavior is one of the classic but obnoxious traits of the child prodigy. Some child prodigies, like Albert Einstein, developed benign personalities. Others, like Mozart, displayed a near intolerable conceit. Unfortunately, Benjamin Gould would be known for his arrogance. Nor was his childhood normal. Gifted children tend to have difficulty relating to playmates, become isolated, and gravitate toward adults early in life. Benjamin Gould's precocity was compounded by an early physical detachment from his parents. As a result, he functioned at an adult level from a very young age. Like many gifted people, Gould also had difficulty in acknowledging authority to which he felt superior, an inability to suffer fools easily. His biting wit and well-known fondness for forceful practical jokes provide two more classic symptoms that modern clinical diagnosis would see as indicative of a disconnected childhood, again typical of child prodigies. But hearkening back to the growth experiences of childhood does not supply a convincing explanation for Gould's subsequent behavior.

Freudian emphasis on childhood experiences has yielded steadily in the past fifteen years to an increasing awareness of the importance of body chemistry and genetic predispositions in emotional disorders. Given the state of theoretical upheaval in psychiatry, the problem of redefining the wide range of behavior encompassed in schizophrenia, for example, presents tremendous difficulties. Schizophrenics and manic-depressives share many symptoms, and the middle ground can be confusing to the diagnostician, bewildering for the layman. Certain personality and behavior patterns, however, do provide enough data for the diagnosis of clearly distinguishable cases. The absence of personal diaries leaves the researcher with few hints as to how Benjamin Gould would fit into currently debated models of schizophrenia. We do know, however, that Gould experienced cyclic periods of elation and despair, alternating periods of high productivity and energy with times in which he was completely incapacitated by what he described as "severe cerebral attacks."

Migraine headaches are incapacitating. But descriptions of migraine normally involve some passing reference to abdominal discomfort, scintillating lights, or pulsing pain in the temples. Gould's letters were filled with anguish during his low periods, but in describing his pain he made no mention of specific physical symptoms other than describing his attacks occasionally as headaches. In other words, he never mentioned a particular kind of pain, such as a pain over the right temple or a shooting pain in the left eye. Instead of listing symptoms associated with migraine, Gould described his headaches as times of exhaustion and complete inability to accomplish any work. His cerebral attack seems to have been a time of disorientation rather than of physical pain centered in a specific location. Disorientation and confusion are typical symptoms of depression; in his letters Gould often described himself as depressed.

At times, Benjamin Gould descended to a level of helplessness and despair that Bache termed "suicidal." The astronomer's lowest points seemed to come mostly in late winter and early spring, still a classic season for depression. During these bad periods, he had trouble sleeping, lost interest in his work, and found little pleasure in life—all symptoms of severe depression. Many of these spans of extended hopelessness had no apparent external cause that would warrant such despair, a fact he often acknowledged himself. In contrast with these psychic ebbs, while he was at the other end of the mood spectrum his letters were filled with elated plans for the future, plans that would lead to unquestioned greatness and the fulfillment of his lofty goals. At such times he was self-absorbed, irritable, restless, and preoccupied with his own correctness, with being more perfect, better than anyone else.

Gould's mood-cycles were more pronounced in his thirties, but lasted through his entire lifetime. Everyone has ups and downs; sadness and elation are universal human experiences. Clinical depressive illness is different. The personality disorder laymen know as manic-depressive has been renamed by modern psychiatry as bipolar affective disorder. Current diagnoses indicate that this is a genetically determined illness, now treated chemically. The illness comes and goes on its own, with no critical connections to life events as factors in the onset of any stage. The personality disorder just becomes increasingly obvious as moods shift from one extreme to another, obscuring intelligence with more aberrant behavior. Judgment becomes so impaired that the patient often endangers his own interests, typically through his overbearing behavior with others. Although the cause of the high or low mood does not have to be connected to a real event, the extreme of either stage is enhanced by what is going on in the real world. As real life gets worse, behavior becomes more extreme in either the elated or depressed stage in response. Real events make depression deeper or elation more euphoric and excited.

Paranoia is also typical among manic-depressives who find themselves in contention with others. Suspicion and distrust are common in both extremes of the mood swing, in elation as well as in depression. In the high periods of productivity, suspicion would fall on anyone who attempted to interfere with "the perfect plan." During periods of depression, suspicion would confirm that the odds were overwhelming because others had conspired against the plan and/or the person. In both the manic and depressive stages, aggressive irritability and cantankerous hostility are common. Hostility is typically expressed as a continuing series of complaints of conspiracy. Such paranoia does not involve hallucinations or hearing voices, but is part of the distorted interpretation of real life events by the manic-depressive person.

The history of Benjamin Gould's emotional peaks and valleys, his manner of dealing with others, and his behavior during the conflict is consonant with current clinical descriptions of what laymen describe as manic-depressive personality disorder. The temperamental "maladjustments" of the child prodigy are also obvious. No definitive diagnosis is possible; nor is one really necessary. But the correlations of Gould's attitudes and behavior with modern clinical descriptions are too strong to be ignored completely.

Gould's early precocity and unquestioned abilities led him and others to expect greatness. His personality, however, was not one that brooked opposition easily. Gould's friends respected his talent but knew that he was not in the habit of weighing his words or dealing easily with those he considered to be his subordinates or inferiors. His sharp wit delighted friends, but his scathing observations and talent for devasting mimicry were sometimes imprudent. Max Weber cautioned that the qualities that make an excellent scholar do not necessarily make a man a leader.[32] Dedication to his own mission to establish a true scientific community in the United States brought this highly intelligent but volatile young astronomer to the Dudley Observatory in Albany.

CHAPTER 5

An Ideal Observatory

The agreement reached at the 1855 meeting of the AAAS changed the direction and significance of the Dudley Observatory. Originally, the observatory had sprung from very simple motives—antebellum enthusiasm for astronomy and local pride. The 1855 decision to establish a research institution equal to the finest observatories in the world moved it into a different class altogether. As far as the worthy citizens of Albany were concerned, the agreement promised increasing international recognition for Albany as a center of science. For the members of the newly created Scientific Council—Bache, Henry Peirce, and Gould—the promised heliometer offered the heady prospect of entering European scientific discussions as peers, not as provincials. Enhanced reputations for American astronomers in particular and increased prestige for American astronomy in general would undoubtedly follow. The Dudley Observatory would also serve as a model for the rest of the United States, a signpost pointing the direction to be taken.

In his enthusiasm for the heliometer, Benjamin Peirce

stressed that only the best was good enough. Blandina Dudley responded by increasing her donation from $6,000 to the full $14,000 estimated as the price for an exceptional instrument. She offered several thousand more to secure "an instrument superior if possible to any in existence," a constant theme in the "great telescope race" of nineteenth-century observatory building. To make certain that work on the heliometer was started as soon as possible, Bache suggested that Gould be sent to Europe immediately. As superintendent of the Coast Survey, Bache expressed his own willingness to suffer the inconveniences of the astronomer's absence in such a good cause.[1]

Gould was delighted. He wrote a long letter to Thomas Olcott, apparently in response to an inquiry from the bank president about the estimated expense of outfitting the observatory and bringing it into working condition. Gould's letter is a relatively concise statement of a professional astronomer's requirements for setting up an American observatory capable of standing comparison with the best in Europe in 1855. The letter is interesting from the perspective of lay support for research in antebellum America. Gould's detailed explanations to Olcott indicate that the Albany trustees were aware of the kind of observatory they were establishing and were willing to underwrite the necessary expense. Because most of Gould's requests were met, except where the astronomer decided otherwise, his shopping list reflects a serious commitment by the trustees to the kinds of professional goals articulated by the members of the Scientific Council.

In describing his ideal observatory to Olcott, Gould acknowledged that his "aspirations have been and are large," then offered a list of instruments "which an astronomer of unlimited means would desire in a thoroughly furnished observatory."[2] He began with a great equatorial telescope, the classic instrument of discovery and the showpiece of any observatory. According to Gould, the best equatorials were made by the firm of Merz and Mahler in Munich, the successors of Joseph Fraunhofer. The astronomer warned Olcott that Albany would probably not be able to afford a refractor equal to those possessed by

Harvard or Pulkova, which were considered to be the best in the world. Achromatic refractors were still enormously expensive in the 1850s. Not only were they difficult to construct, but the flint glass industry had been tightly controlled by a few secretive European suppliers until 1848. Import duties remained high as well, pushing the price to prohibitive levels.[3]

Next came the heliometer, the pivotal psychological point for the Dudley Observatory. In describing the Oxford heliometer, George Airy, the Astronomer Royal, emphasized that the instrument was a very poor "seeing" telescope but as a "*measuring* telescope,—the light in which it should be considered,—it is all that can be desired."[4] The prospect of installing such an instrument in the United States had been an irresistible attraction to Bache and Peirce. For their part, the trustees' willingness to underwrite this level of equipment indicated that their intentions were not frivolous. Gould described the heliometer carefully to Olcott, emphasizing that it was designed to measure distances and the relative positions of celestial bodies precisely. It is clear that Thomas Olcott and other trustees had a layman's understanding of the specialized purpose and function of the instrument in 1855.

A meridian circle would be the first telescope to stand in the Albany observatory. This instrument, which measured right ascensions and declinations, was rapidly becoming the workhorse of European observatories, especially in Germany. The Dudley Observatory's Meridian Circle, which cost about five thousand dollars, was long considered to be an outstanding example of optical craftsmanship.[5] Because the building included a room specifically designed for a transit, Gould suggested that the Coast Survey purchase the instrument, mount it in the Albany observatory, then sell it later to the trustees if it were really needed. After discussing other equipment, including an elaborate new clock system, Gould concluded with a statement that would be recalled later by the trustees: ". . . and it is not impossible that astronomy might lose rather than gain if noble struments should be suffered to lie unemployed. But I do not believe this could happen in a city like yours, not only overflowing with opulent, liberal and public spirited citizens and with patriotic and munificent ladies, but also the capital of the Empire State. . . ."[6]

In September of 1855 matters at the Dudley Observatory seemed to be moving forward quickly. The happy consensus was that the observatory would be functioning by August of 1856, when Albany would proudly host the next meeting of the American Association for the Advancement of Science. Gould made plans to sail immediately for Europe. Bache wrote Olcott that the Coast Survey would "cheerfully contribute" Gould's time; the trustees should pay all of the astronomer's expenses, estimated at about $1,200. As Gould hurried his preparations, he rejected Bache's suggestion that he visit Albany before departing, not wishing to spare the time. Bache's advice was both politic and practical. Gould's habit of ignoring the "niceties" as well as his habitual secrecy in his dealings with the trustees were first intimated here.[7]

Gould carried authorizations that empowered him to purchase any instruments he might think necessary for the observatory. The trustees themselves advanced the sums required for deposits and personally guaranteed the remainder of the purchase price for everything he ordered. All of these expenditures were made out of their own pockets. As Gould sailed in late September of 1855, Peirce wrote confidentially to Bache of Gould's "golden opportunity," saying "if he fails now, all is lost." Bache hoped to insure against failure. Within a few days, he sent a formal letter of agreement to the trustees, explicitly restating his offer to provide instruments, including a transit, as well as an observer from the Coast Survey "if the means to purchase a Heliometer" were provided.[8]

While Bache was locking up the terms of agreement, Gould was elated at the prospect of returning to Europe. His euphoria carried him in high spirits through the rough gales of his twelve-day crossing. Letters written to the Astronomer Royal from on board ship prophesied a brilliant future for American astronomy, with promises of useful observations flowing from Albany by October of 1857. Gould also asked Airy to write to Mrs. Dudley, saying that a few lines from him would "be the source of extreme gratification to the good old lady."[9] The cordiality of Gould's letters to Airy provides an interesting contrast with the consistently derogatory references to the Astronomer Royal in personal correspondence between Gould and Peirce. These cutting remarks might have had their origin in American resentment of their position on the scientific periphery while Airy took up so much room in the middle. Or the animosity may have had its source in Gould's aversion to English methods in astronomy or to Airy's authoritarian personality. Whatever the basis, Gould's letters to the Astronomer Royal were warm and cordial. His scathing observations about Airy were reserved to his private correspondence with fellow Lazzaroni.

Gould had only six weeks' leave from the Coast Survey. Ordering so many new instruments, each of which was to be "the best available," required time-consuming consultations. He realized he would have to move as quickly as possible from Paris to Berlin and Munich, returning by way of Oxford to examine that heliometer. Paris was overflowing with people when he arrived. The Great Exhibition of 1855—an industrial display of unprecedented magnitude—attracted visitors from all of Europe. The astronomical instruments displayed there were truly splendid in Gould's estimation. He renewed his acquaintance with Charles Babbage, creator of automatic calculating machines now recognized as direct precursors of the modern computer. Such machines were of great interest to astronomers. Preparing tables of astronomical positions was sheer drudgery. On greeting Gould, Babbage immediately launched into a description of an "analytical engine" developed by George Scheutz and his son, which was on display at the exhibition. The Scheutz Difference Engine, influenced by Babbage's earlier designs, calculated complex tables using the method of finite differences. The machine also produced a mold from which tables could be printed. Gould purchased the Scheutz Difference Engine for the Dudley Observatory, where it remained until the early twentieth century. The Scheutz machine is now part of the national collection of the Smithsonian Institution.[10]

Not all of Gould's time was spent so seriously. He arranged to meet old friends, the Honorable Josiah Quincy and his family, and spent several days strolling the streets of Paris with his future wife and in-laws. Urbain Leverrier (1811–1877), the French astronomer, gave a party in Gould's honor, inviting all the astronomers who were in Paris for the exhibition. The gathering was full of enthusiasm for the new observatory in Albany. Babinet offered to write an article for the *Revue des Deux Mondes;* Babbage promised a personal letter to Mrs. Dudley. Later in the week, Gould conferred with Leverrier about technical innovations in the instruments to be ordered. Summing up his experiences to Bache, Gould wrote that preparations for the observatory were proceeding splendidly, describing his week in Paris as one that had compressed more joy and grief than any other in his life.

On 26 October Gould signed a contract with Pistor and Martins of Berlin for the meridian circle. The contract specified that the circle would "be equal in every respect to that constructed by them for the University of Michigan at Ann Arbor" and was to be tested until it satisfied Professor Encke of the Berlin Observatory. Gould also engaged the same firm to construct the Coast Survey transit for which the trustees provided the initial deposit. Not everything went so easily. The heliometer proved especially difficult. The Hamburg firm of Repsold, makers of the Oxford heliometer, refused to construct an object glass larger than nine inches. They were equally adamant in their refusal to be hurried, enforcing their reluctance with a prohibitive price. Gould decided to look elsewhere. The Munich firm of Merz and Mahler proved equally frustrating over a refractor. Gould confided that his dealings with that company had "greatly diminished my respect for the firm, in their business relations."[11] He was more successful in contracting for the observatory clock and left a deposit with a craftsman in Altona. Although the clock system was to be donated by Erastus Corning, the deposit was advanced by the trustees because Corning refused to pay his subscription until the clock actually arrived in Albany.

Leaving Germany, Gould traveled to the Greenwich Observatory. During this visit Airy advised him to insist on an endowment of one hundred fifty thousand dollars for the Albany observatory. At Gould's request Airy wrote to Mrs. Dudley in December of 1855. In commending her generosity in funding the purchase of a heliometer, Airy emphasized the importance of the work to be done with the instrument but returned to the theme he had stressed to Gould—an endowment to secure the future of the observatory. The Astronomer Royal stressed that work did not finish once the equipment had been set up and the building admired. No science required more follow-up work than astronomy, as observations were filtered through calculations combining past, present, and future. Because the United States was a young country, Airy cautioned her that Americans usually found it easier to think of the present rather than make provision for the future.[12]

Lack of an endowment had closed the Cincinnati Observatory and sent its astronomer out to work on the railroad, despite the excellence of the refractor standing in the building. The endowment of the Harvard College Observatory was sufficient to maintain both the observatory and astronomer, but there was not enough income to heat the observatory library and computation rooms during the winters in which George Bond painfully declined with tuberculosis. Nineteenth-century philanthropists could easily be persuaded to build an observatory, outfitting it with the best instruments available, especially if the institution would bear the family name. Purchasing instruments and installing them in classical buildings was far more satisfying than setting up a source of interest-bearing capital to pay an astronomer to do his work at night. Endowments were less visible status symbols than buildings and telescopes.

Leaving Greenwich, Gould stopped by Oxford to examine that heliometer, then sailed for New York, suffering an equally stormy return voyage. He arrived in Boston in a raging snowstorm on 30 December, allowed himself two days at home in Cambridge, then set off to Albany. After a brief visit, Gould and a few friends traveled on to Canastota, New York, to see Charles A. Spencer, a self-trained optical craftsman. Spencer had achieved enormous success in creating the first American microscopes capable of competing with the best European-made instruments.

Spencer and his partner, A. K. Eaton, were part of a growing community of American optical craftsmen. As instrument makers increased their skills at mid-century, the American market for optical instruments also changed, but not uniformly. Nationalism was strong with telescope construction because observatories were large-scale community efforts. The Harvard refractor of 1843, built by Merz and Mahler, would be the last of the great refractors built by Europeans for American use. In contrast, microscopes were purchased by individuals. Many young physicians who returned from European study preferred a familiar German or French microscope to an unknown American instrument. As Spencer's instruments began to compete favorably with the best European microscopes, however, national pride overcame earlier reluctance. Canastota became the "microscopic Mecca," a place of pilgrimage for the leading American scientists and physicians of the 1850s.[13]

Gould later maintained that Dr. Armsby was the first to suggest that Spencer construct the heliometer, but the astronomer was equally enthusiastic at the time. While visiting Spencer, Gould was impressed with the thirteen-and-a-half-inch equatorial telescope the craftsman had built for Hamilton College in 1855, the largest telescope built in the United States up to this time. Although this was Spencer's very first telescope, Gould declared that the mounting was equal to Munich-made instruments. In praising Spencer's skill to the Astronomer Royal, Gould wrote that he had "taken a step which will probably bring upon me a considerable ridicule in Europe until the instrument is completed at which time I trust that the laugh may be turned."[14] Gould asked the Canastota craftsman to build the heliometer for the Dudley Observatory. Spencer eagerly accepted the challenge. He suggested modifications in Repsold's designs for the Oxford heliometer and promised to introduce innovations that "combined the advantages of Bessel's divided object-glass and . . . [Airy's] double image micrometer."[15]

Spencer also agreed to make a comet-seeker for the observatory. Unfortunately, Charles Spencer had a well-deserved reputation for being notoriously slow in his work.

Gould and his party returned to Albany in high spirits. A gathering met at the home of Mrs. Dudley on 4 January 1856 to hear Gould describe his successes and failures with European craftsmen. The Astronomer Royal's letter to Mrs. Dudley on the endowment was read aloud by one of her relatives. Gould then made the dramatic announcement that the heliometer would be built by an American—Charles A. Spencer. He emphasized that an American-built heliometer in a great American observatory would be an inspiring source of national pride in American scientific achievements for both scientists and their patrons.[16]

At a meeting held at Judge Amasa J. Parker's home several evenings later, Gould urged the trustees to send Spencer to Europe to study construction methods used in the best European optical factories. Gould suggested that John Gavit, a friend of Gould's, should be sent on the journey also. Spencer was not used to traveling, and Gould insisted that "his life was too precious to the cause of science to be risked." The trustees were persuaded. They agreed to pay for a second European tour, for two this time, on Gould's recommendation.[17]

Although Spencer confessed that he felt at a disadvantage because he had never seen a heliometer, much less made one, he was confident of his ability to fulfill his promise. He wrote Thomas Olcott, giving estimates of the probable cost. Gould and Spencer had discussed three possible sizes for the object glass, weighing the advantages and difficulties in glasses of eight-, nine-, or ten-inch diameters. The Repsolds had quoted a price of $9,000 for a seven-inch, $11,000 for an eight-inch, and $13,250 for a nine-inch object glass. Spencer and his partner declared themselves ready to build a heliometer for the Dudley Observatory with an eight-inch diameter for $7,500, or a nine-inch object glass for $10,750. No effort would be spared, he said, because "the successful execution of such a work would be an additional gratification to Mrs. Dudley, if accomplished by American skill."[18] Although the trustees had been persuaded to underwrite Spencer's European trip, they later decided that the costs should be applied to his estimate for the heliometer. Interest on the deposited donations had brought the total available in the heliometer account to $15,033.22. What is clear from the minutes of this meeting of the trustees is that, however reluctantly, they were using capital, not interest, to pay travel expenses and deposits. In January of 1856 the trustees were placing their trust in Gould's leadership, following him down a path they hoped would bring the observatory into operation as quickly as possible.

As preparations for Spencer and Gavit's departure moved along, Gould suggested that they also purchase a standard barometer and special thermometers, which would have to be carried by hand on their journey. Unless personally carried, British import regulations required the instruments to be removed from their original packing cases, repacked, and shipped across England by British transport before they could be sent on to the United States. Neither the maker nor the insurer would reimburse losses suffered in the process. Gould promised to give Spencer detailed instructions about proper handling, reassuring the trustees that Gavit would see that they arrived safely in Albany. Broken barometers and thermometers would later be added to a growing list of irritations between the trustees and Gould.

Gould remained in Albany during January, attending parties and meetings. Bache spoke in the Young Men's Association lecture series, and the two scientists were beseiged with invitations. Gould and Bache contacted several railroads, including the New York Central, the Hudson River Railroad, and the Western Railroad—offering to sell accurate time from the Dudley Observatory. An agreement reached with the New York Central Railroad promised to provide an income of fifteen hundred dollars per year as soon as the observatory could be brought into operation. Gould also negotiated with Springfield, Worcester, Boston, Utica, Rochester, Buffalo, and New York City about providing accurate time by telegraphic signal. The time signal was to be transmitted by a method described to Gould by Airy. A "time-ball" apparatus dropped at noon in London every day, activated by a telegraphic signal from the Greenwich Observatory. Bache drafted a proposal for a similar connection between Albany and New York City. The trustees hoped that this service would encourage the merchants of Manhattan to make contributions to the observatory's endowment. By September of 1856 it was apparent that the Manhattan connection was defeated by local jealousies. A time-ball connection between the Dudley Observatory and New York City would eventually be made in 1860.[19]

The trustees were delighted with the agreements reached. Providing accurate time to railroads and cities served a dual purpose. Once it was operational, the Dudley Observatory would be able to contribute toward its own support, allowing the endowment to accumulate interest. Selling time would also establish the new institution in the public mind as a working observatory, gaining credibility and more donations while it was readying itself for its more serious purpose. With so many costly inroads into capital from travel expenses and deposits on instruments, the trustees were anxious to bring the observatory into operation as quickly as possible to increase public support. The efforts Bache and Gould made to bring this about increased the trustees' confidence in the commitment of the members of the Scientific Council to the observatory. In response, the trustees were willing to extend themselves beyond their normal cautiousness. These decisions were based on a mutuality of purpose and a sense of trust. In January of 1856 matters in Albany seemed to be going exceedingly well. The trustees were pleased with the arrangements with the railroads. Spencer and Gavit were preparing to leave for Europe. Gould left for New Orleans to take up his Coast Survey duties.

Gould worked with his Coast Survey field parties for the next two months, corresponding with Dr. Armsby about observatory details as they arose—instructions for Spencer and Gavit, limestone piers for the telescopes, invitation lists for the inauguration of the observatory, the upcoming AAAS meeting in August, and so on. He returned to Cambridge in April to begin reducing the observations on Mars and Venus made during the Gilliss Expedition to Chile. He spent four weeks on that work and on his own project of determining the personal equation in telegraphically signaled observations, but Albany was demanding increasing amounts of his time. Preparation for the tenth meeting of the AAAS as well as the simultaneous inauguration of the Dudley Observatory and the new Geological Hall that would house the Natural History Survey's collections soon precluded anything else. Unfortunately, Gould's mood entered a period of decline. He complained of the "distracting draughts upon my presence and thought and time and nerves" caused by the flurry of activity in Albany.[20]

Gould was general secretary of the AAAS at this time, in charge of the arrangements for the meeting. Dr. Armsby was chairman of the local committee. Inordinate amounts of Gould's time, thought, and correspondence were dedicated to deciding who among the most eminent European scientists would be offered a free passage across the Atlantic to attend the meeting. Gould insisted that only "A-1 scientists" from Europe should be asked, and the invitations were to be personally offered in declining order of importance by Spencer and Gavit. Gould was not the only one who worried that the invitations might be given out indiscriminately. Louis Agassiz asked James Hall to tell the two travelers "for mercy sake not to extend the invitations to third rate men, even if the best do not come." They were equally discriminating with regard to choosing which American scientists would be given predominant roles. Gould and Agassiz did not want to give "our tenth rate creepers . . . too good a chance to make themselves big." Dr. Armsby, James Hall, Benjamin Peirce, Louis Agassiz, and Gould conferred frequently as they put the program for the meeting together. Armsby invited political figures in New York, although Gould worried that he was "going off the track" in doing this. Gould negotiated delicately with Edward Everett—former congressman, ambassador, president of Harvard, and the leading American orator of the nineteenth century—who placed Gould in an embarrassing position by asking to speak on astronomy on both days of the celebration.

Benjamin Peirce finally persuaded Everett that it was only appropriate to allow the geologists to address the opening of the Geological Hall on at least one of the two days.[21]

If the Dudley Observatory were to be ready for the August 1856 AAAS meeting, the building demanded extensive preparations. Gould wrote Professor Challis of the Cambridge Observatory in England that matters in Albany were not progressing very rapidly, but he hoped that "what is lost in time may be gained in quality."[22] Why was the Dudley Observatory taking shape so slowly? The Ann Arbor Observatory had been started in 1854 and was progressing rapidly, a sore point with the citizens of Albany. One obvious cause of delay was Gould's insistence on remaining in Cambridge. Gould continued to direct "minutest particulars" of alterations, but he did it all from his home in Cambridge. Gould put an enormous amount of time into the effort to bring the Dudley Observatory into shape, but it was his spare time. He never assumed the title of director during this period; he never stayed in Albany for more than one or two days. In May of 1856 Gould and Bache decided against Gould's taking official charge of the Dudley Observatory. Bache believed it would be injudicious for Gould to leave the Coast Survey unless there was some certainty that the observatory could continue on its own if someone besides Bache should become superintendent. Again, it was the thorny question of the endowment. Gould attempted to soften the blow by writing Armsby that he was devoting all his energies to equipping the observatory.[23]

The itinerant career of O. M. Mitchel and the impoverished Cincinnati Observatory had no appeal to Benjamin Gould. He had no desire to attach his own name and reputuation as a "professional astronomer" to the Dudley Observatory if its future were to be as insecure as the Cincinnati institution's had proven. Gould had a strong aversion to begging for money for the support of science, even for his own beloved *Astronomical Journal*. In early August, as the ceremonies approached, Benjamin Peirce wrote Thomas Olcott, stressing the importance of an endowment that would yield a minimum of ten thousand dollars per year. Rather than attempt to run the observatory for less money,

Peirce advised that the instruments remain idle until enough funds were accumulated.[24]

It is obvious that an incipient disagreement between the trustees and the scientists was in the offing on this question. The scientists were understandably wary of voluntary funding for such an important enterprise. Bache, Peirce, and Joseph Henry preferred the security of public support. As far as the trustees were concerned, they could not understand why it was preferable to build and equip an observatory with an endowment of reasonable size within American standards, and then let it remain idle rather than enabling it to fund itself by providing time to railroads and merchants. The fifteen hundred dollar yearly contract with the New York Central as well as other cities provided an incentive in the trustees' minds to bring the observatory rapidly into operation. The trustees reasoned that a working observatory would encourage contributions to the endowment more than one that was closed. This conflict in perspective would become increasingly important.

In a letter to Olcott in early August, the scientists played deftly upon the heightened expectations and anticipatory pride of the trustees as the inauguration ceremonies approached. Stressing the observatory's importance to the scientific future of the United States, they promised it would be "second to none in the world" if a yearly income of $10,000–$12,000 were provided. Bache, Peirce, Henry, and Gould urged the trustees to solicit financial support from all parts of the state, especially New York City, where there was talk of an attempt to establish a rival university and observatory that the scientists wished to discourage. They stressed that no other observatory in the United States could show such excellence in its instrumentation and planning as the Dudley Observatory.

In the meantime Gould concentrated on structural alterations. The meridian circle was to arrive first, so he gave his attention to the east wing, which would house that telescope. Park-like settings and architecturally interesting scenes near urban areas were important to the "gentlemanly" quality of life in the twenty years before the Civil War, as the rural cemetery movement demonstrates. Because the observatory was a scenic part of Albany, Gould's east wing addition disturbed some who wished to maintain the symmetry of the building's classic design. After discussion, Gould agreed to a simultaneous alteration of the west wing. The rear walls of both wings of the building were torn down in July of 1856. Insertion of the huge limestone piers that Gould stipulated for the instruments would, of course, have necessitated breaking down the west walls at some time anyway. The trustees reasoned that if the work were done at the same time, the building would have two balanced additions, but Gould resented the trustees', mainly Dr. Armsby's, interference in this matter.[25]

The ever-busy Armsby was obviously immersing himself completely in the alterations, as well as in the preparations for the inauguration ceremonies and the AAAS meeting. Armsby lived in the medical college he had helped originate, was closely involved in the city hospital that had arisen through his efforts and those of his father-in-law, Dr. Alden March. He took an active and continuing interest in the law school he "saved" in 1854. It is not surprising that he took such a personal interest in the many institutions he had brought into being. The joys of setting out trees, adding rose trellises, shoring up eroding banks, and having opinions about absolutely everything that was going on were part and parcel of the same impulses that led James Armsby to circulate the first petition to establish an observatory in Albany. The energetic physician had been instrumental in securing the original charter of the Dudley Observatory from the state legislature and could say, in all honesty, that from the very beginning through the decision to purchase a heliometer, the observatory had been "his" idea.

Armsby was later described by Gould as "busy, untiring, gossiping, and wondrously meddlesome."[26] His continual suggestions were understandably resented by Gould, who felt, with justification, that many of the doctor's ideas were irrelevant to the scientific purposes of the institution. Gould had little interest in how the observatory looked. Armsby's aggravating "meddlesomeness" was unquestionably irritating, but "the busybody" had produced beneficial results for the city of Albany. However annoying he might be, Armsby knew how to mobilize support locally in order to get things done. Unquestionably, Dr. Armsby was a difficult man, but he was better tolerated than alienated. In retrospect, given these two personalities, conflict between the physician and the astronomer seems almost inevitable.

Gould's insistence on detail, his "excessive particularity" as it was later described, and his demand for complete control were directed toward equally beneficial ends—a reputation of scientific excellence for himself and the Dudley Observatory. But Gould's insistence on exercising that control from Cambridge, his refusal to explain the reasons for seemingly overparticular requirements, his brusque rejection of suggestions—all combined to give an impression of arrogance and insensitivity. The observatory was not yet in working condition. The coming gathering of American scientists would thrust the observatory into the national spotlight; the trustees were most anxious that the institution make a positive impression to establish its importance and encourage support. Yet their efforts to help infuriated the astronomer in Cambridge.

Gould's position was an ambivalent one in 1856. He refused to assume an official and therefore a publicly responsible connection with the Dudley Observatory, insisting that he served only in an advisory capacity. The trustees had reluctantly accepted this arrangement. Such a role conferred authoritative weight, but it also required sensitivity and tact because of its tenuous character. Gould's haughty arrogance—the legacy of a child prodigy—made for poor politics in an institution in which his own role was so undefined. In contrast, Bache and Henry aspired equally to scientific excellence in their respective institutions—the Coast Survey and the Smithsonian—but they also accepted real responsibilities. They were conscious of the need to deal courteously and carefully with those who provided essential financial support. The Bonds at Harvard Observatory were equally adept at maintaining amicable relations with their visiting committees. Gould had a small constituency—a group of already willing and supportive trustees. He lacked both the reputation and stature of Bache or Joseph Henry and the institutional stability of a university such as Harvard. Yet he attempted to wield authority in a manner that antagonized the men who provided the very basis of his position, in effect working against his own best interests.

When discussing the alterations going on at the observatory during 1856, Olcott later stated that the building committee was unable to assist in any way without offending Gould. As frustrations built on both sides, Olcott finally told the workmen on the site to take all their orders from Gould alone. They were not to drive a nail without approval from Cambridge. Gould clearly resented any attempt on the part of the trustees to speed up the remodeling of the observatory; but he refused to supervise it himself on site. Workmen waited on full pay for answers to their questions to arrive by letter from Cambridge. Despite these restrictions imposed by Olcott, Gould complained that his orders were disregarded during his absence, "a disregard which convinced me, at an early date, that the empty dazzle of temporary show was, in the wishes of the managing Trustees, paramount to any ideas of scientific usefulness or dignity."[27]

For the trustees, "empty show" served a specific purpose. Contributions depended upon the public perception of the Dudley Observatory as a cause worthy of hard-earned dollars. Newspapers throughout the United States would carry stories about the meeting of the "scientific savants" in Albany. Those stories would include discussions of the prospects for the observatory's success. Gould's refusal to recognize the practical reasons behind the trustees' desire for an impressive physical facility at the August meeting was not realistic in relation to his own demands for an endowment larger than that of any other comparable institution in antebellum America.

During the summer Gould made another decision that would plague him in the future, but it seemed of little importance in July of 1856. Because Gould intended to remain in Cambridge, he decided to send a representative to Albany. Gould had hesitated to delegate any authority because he believed assistants left on their own tended to "grow in their own estimation and assume credit for work done under another's direction," but he changed his mind. Dr. C. H. F. Peters, the European astronomer who had extolled the merits of a heliometer to Peirce, would supervise construction at the observatory while continuing his work for the Coast Survey. Gould praised Dr. Peters as "a man of a good deal of experience and excellent attainments, and I [Gould] have only hesitated on the matter of policy." From Cambridge, Gould urged speed on both Peters and Armsby in bringing the observatory into good condition for the inauguration in August, one month later. Dr. Peters was a competent astronomer, expert in the German methods valued so highly by Gould. More importantly, Peters made an extremely favorable impression on the trustees. Olcott described him as a gentleman, unassuming, truthful, practical, and progressive. Peters arrived as a foreigner in July of 1856, but within a very short time made influential friends throughout the city, including James Hall and Stephen Van Rensselaer IV. Peters set to work at once, making observations in the midst of the chaos of workmen and remodeling at the observatory.[28]

As August grew closer, the city erected tents in Academy Park, and the *Albany Atlas and Argus* described the upcoming meeting as the "greatest assemblage of its character that ever convened this side of the Atlantic." Great credit was given to the work of Dr. Armsby and his untiring exertions in arranging entertainment and accomodations for all the visitors. The meeting of the AAAS opened on the morning of 20 August 1856 with full pomp and ceremony. Despite the offer of free passage across the Atlantic, no European scientists attended. The first day's meeting was marred by a heated dispute arising from Bache's attempt to centralize and strengthen the powers of the Standing Committee. Many members felt that the influence of Bache's friends was already too powerful on the committee and, when the issue was raised, "a warm opposition appeared," led by William Barton Rogers.[29]

Meetings continued through the week, reported at length in the *New York Times* and the local Albany press. At one of the smaller sessions, Gould described the meridian circle and the transit ordered for Dudley Observatory. In his explanation Gould emphasized his preferences for German styles of instrumentation, characterizing English instruments as appropriate to the engineer and the German to the artist. German instruments were lighter, mobile, versatile, and accurate; in contrast, the English stressed uniformity, he declared. As he concluded his paper, Gould announced that the trustees had named the observatory's meridian circle in honor of Thomas W. Olcott and had ordered the banker's name engraved upon the telescope.[30]

With the business of the AAAS complete, the ceremonies for the dedication of the Geological Hall were held on 27 August. The honorary platform bulged with eminent politicians, ranging from former president Millard Fillmore to assorted governors, present and past, as well as respected scientists like Louis Agassiz and James Hall. Hall had commented earlier to John Fries Frazer that the occasion offered a splendid "opportunity for saying many good things about science" and the chance would not occur again in their lifetimes.[31] After all the speeches, the Geological Hall was duly dedicated, and the entire association adjourned to a reception held that evening at Mrs. Dudley's home.

The observatory was to be dedicated the next day. Despite all the grand plans, the appearance of the Dudley Observatory in August of 1856 was not one that inspired confidence. The trustees later described their embarrassment during the inauguration ceremonies: "The Observatory itself had the appearance of a ruin. The walls of both wings were open to receive the piers and cap-stones, and to permit the workings of the "Ingenious Crane," . . . instead of leading their distinguished visitors up the hill, to spread before them the glories of the model instruments, the efforts of the Trustees were directed toward keeping them within the limits of the Capitol."[32]

The great day finally arrived, sunny yet cool, with five thousand people seated as the ceremonies began. Former Governor Washington Hunt gave a eulogy of Charles E. Dudley. Gould spoke next. Although engaging and witty in private conversation, Gould experienced paroxyms of anxiety when faced with large groups.[33] Briefly recounting the fruitless attempt to establish a national university in Albany in 1852, Gould reviewed the early history of the observatory. Then, in remarks that must have haunted him later, the astronomer heaped lavish praise upon Dr. Armsby. Armsby received full credit for the change in fortune of the Dudley Observatory and was described by Gould as a "man whom to know is to love, and to mention is to praise. . . God bless him! for he is blessing God's earth, and the world is better that he lives in it."[34]

Bache, to whom public speaking came more easily, rose next to praise Gould's abilities and to report that twelve citizens of Albany had contributed the sum necessary to support Gould's *Astronomical Journal* for the next six years in connection with the Dudley Observatory. The twelve donors had been organized by the indefatigable doctor, James H. Armsby. Judge Ira Harris then read a letter from Mrs. Dudley announcing her decision to contribute an additional fifty thousand dollars to the endowment. Although she had decided to do this earlier, her decision had been kept secret. The announcement was greeted with uproarious applause from the audience, and Agassiz led the assembly in three cheers in honor of Mrs. Dudley by swinging his hat.

The credit for persuading Mrs. Dudley to increase her donation would become another petty point of contention between the trustees and the members of the Scientific Council. The trustees insisted that Thomas Olcott had been the persuasive voice; the Scientific Council insisted the donation was a direct response to a letter they had written in August. Apparently, the widow was considered so pliable that the credit of "who asked?" became the credit of securing the gift. Mrs. Dudley's competency to handle her own affairs would be another issue in the controversy as both sides attempted to use her name to control the institution.

Blandina Bleecker Dudley was descended from some of the most respected and influential families in Albany's long history. She inherited great wealth, married a successful man, and dispensed money liberally to charities within the city. During this period of the observatory's relation with the Scientific Council,

Thomas Olcott was her closest advisor. She had taken control of her affairs away from her lawyer, Harmon Pumpelly, a close associate of Erastus Corning's, and placed the management of her property in Olcott's hands. In May of 1856 Mrs. Dudley became estranged from her nephew and principal heir, Rutger Bleecker Miller. After one particularly acrimonious quarrel, she spoke to Olcott about drafting a new will. Olcott suggested in June that she consider givine one third of her estate to the Dudley Observatory. It is clear from the Olcott papers that Mrs. Dudley and Thomas Olcott had extensive discussions about transferring a large sum to the observatory before the August letter from the Scientific Council arrived. The letter from the scientists no doubt inspired her with the dramatic gesture that would add to the excitement of the inauguration ceremonies. Rather than have the funds come later through her will, when she would not be around to enjoy the results, she could give the money now and hear the applause.

The donation to the Dudley Observatory endowment was made during this period of estrangement between Blandina Dudley and her nephew. Miller was appalled to see this large amount move out of the family estate and took immediate steps to gain control of her affairs. Miller wrote Olcott in December of 1856 to advise him that "under a painful sense of duty . . . the heirs of Mrs. Dudley owe it to her as well as to themselves, to take some decisive measure, as to the future management and disposition of her Estate." He insisted that after observing her, he had concluded that she was incompetent and could not be considered responsible. He described her mental condition as "incoherent, inconsequent, and oblivious" and concluded that, in his judgment, "some immediate action is necessary." By April of 1857 control of all of Mrs. Dudley's interests rested with Rutger Miller. These efforts have an important bearing on the use of her name and influence later in the controversy.[35]

As the quarrel between the scientists and trustess later developed, Rutger Miller spoke publicly for Mrs. Dudley, attacking the trustees. He worked vigorously to try to get the charter of the observatory overturned, continually attempting to withdraw his aunt's donation to the endowment. Mrs. Dudley's mental competency and the use of her name by her heir in an effort to rescind her donation are just one of the local factors entering as variables in this controversy. Was Mrs. Dudley competent when she made the initial donation to the observatory in memory of her husband? Was she rational when she donated the sum for the heliometer? Was she sane a year later when she contributed fifty thousand dollars to the endowment? Four months after the inauguration ceremonies, Miller did not think so. Yet one year later, both Miller and the members of the Scientific Council would be arguing that Senator Dudley's widow was fully competent when she, through Miller, demanded the resignation of the majority of the trustees from the board.

Whatever her mental abilities, Mrs. Dudley was the center of attention at the inauguration ceremonies for the Dudley Observatory because of the liberality of her gift. The dedication speech for the observatory was given by the renowned Edward Everett, who spoke for two hours, entirely from memory. Although he later described the speech as a flabby set of platitudes, Gould confessed that during Everett's delivery, he found himself, "intoxicated . . . bewitched and carried captive" by the oratory.[36] Following the stirring ovation that met Everett's conclusion, the crowd called again for Mrs. Dudley. She rose and bowed, shedding tears of deep emotion, overcome by the compliment.

The AAAS meeting was over. The general consensus was that it had been a "most brilliant affair." Dana wrote James Hall that Albany "did nobly indeed," especially complimenting the "untiring efforts of Doctor Armsby." As one enthusiast wrote William Bond of the Harvard Observatory, "This Dudley Observatory is likely to be a fine thing. What a passion on this subject is spreading!" Bache and Gould adjourned to Mount Desert in Maine, where they spent their time "talking Albany, building 'astles in the cair [*sic*] and surveying the Coast."[37]

During the fall memories of the inauguration ceremonies provided a continuing glow of pride in Albany. Copies of the speeches were bound as a pamphlet and distributed. The observatory was described in newspapers throughout the United States as a fitting American example of the liberality that could be shown by an enlightened citizenry in a democracy. Patronage of science in the United States could spring from the disinterested benevolence of citizens rather than depending on the whims of autocratic rulers, according to several editors. With such warm endorsements, both the trustees and Gould were hopeful of bolstering the endowment. The trustees increased their own contributions substantially and hoped that others would follow their example.[38]

Despite the elation, delays persisted. The stone piers for the meridian circle and transit were overdue. The dimensions of the great blocks had been set by Gould, after long lithologic discussions with James Hall during the spring and summer of 1856. Gould hoped to mount the most stable and accurate instruments in the world at the Dudley Observatory. No stones of comparable size have ever been ordered since as a foundation for any telescope. Despite the great care in specifying exactly what was needed, flaws in the stones were spotted by Dr. Peters when they finally arrived. Not only were the piers traversed by longitudinal cracks, but a piece of one of the corners had been broken off and patched in an attempt to conceal the defect. Peters asked Gould, who had returned to Cambridge, to come to Albany immediately. Had Gould been present at the observatory when the stones arrived, he would have rejected them and been able to explain his reasons to the trustees. An incident that resulted in frustration and ill will on both sides might have been avoided.[39]

Peters had not only written Gould, he had also advised Dr. Armsby of the problem. Armsby obviously had little understanding of the technical requirements that were of such overwhelming importance to Gould. As he peered closely at the stones, the doctor decided that the cracks seemed barely noticeable. Taking far too much upon himself, Armsby irresponsibly told Peters not to bother Gould any further and simply have the stones put in their proper places. Gould had, after all, urged

Armsby to see that the observatory progressed as rapidly as possible. In the meantime Peters waited anxiously for word from Cambridge. While the letters passed in the mail between Albany and Cambridge, a bill for the stones was submitted by the quarrymen and an unaware Olcott paid it. It was, after all, the normal custom to pay for goods immediately upon delivery. Several days later the answer came from Cambridge. Gould declared himself outraged at the attempt to conceal the chip on the corner. The astronomer would later charge that Armsby himself ordered the concealment of the defects, but this was never proven or accepted locally. By the time such charges were being hurled, neither side could be called unprejudiced. Gould's information seems to have come from the Lockport quarrymen who supplied the stones, whose motives in disclaiming fraudulent participation and blaming Armsby are obvious.[40]

Armsby's interference was unquestionably irresponsible. But Gould's reticence to assume official responsibility is an underlying cause of such problems. Trustees were asked for opinions and immediate decisions. Bills were submitted, and the one man who had to be satisfied was miles away in Cambridge. The situation was guaranteed to create difficulties. Only the maintenance of trust on both sides could smooth things over. Gould refused to use the stones. The quarry refused to take them back. The rejected monoliths stood on the observatory grounds for years and were referred to by the trustees as a "monument of our own folly in yielding to excessive particularity."[41] Gould's refusal to explain himself or his requirements was baffling to the trustees. The trustees, who had paid for the stones, were now in the position of trying to secure new ones. Gould's seemingly arbitrary rejection meant more delay, disappointing everyone. The entire incident—reflecting the poor communication between the nominal director and the trustees—boded ill for the future. As "unofficial" scientific director of the observatory, Gould enjoyed a considerable amount of power without the attendant responsibilities. Expenditures were made at his order because he was trusted. Should that trust be destroyed, the expenditures would be examined and become a source of recrimination.

The obvious delays in the work at the Albany observatory gave rise to fear for its future among the members of the Scientific Council as well as among the trustees. After the excitement of the inauguration, the molasses-like pace of the observatory's progress prompted Bache to ask Peirce, "Whether our Albany friends can carry it or no. . . . I hope we may live to see it but Who knows?"[42] In late November of 1856 Joseph Henry, Bache, and Olcott spent several days in New York City in an effort to secure contributions to the endowment from Manhattan merchants. The fruitless struggle led Bache to reevaluate the Albany situation. If the Dudley Observatory did not quickly assume its own self-sustaining momentum, the members of the Scientific Council would quietly drop it and move on to more promising opportunities for American science. They had no commitment to the institution itself, only as a means to a larger end. In contrast, for the trustees, the Dudley Observatory was an end in itself.[43]

In December of 1856, four months after the high praises of the inaugurating speeches, Gould was actively pursuing a professorship in mathematics and astronomy at Columbia University, strongly supported by Bache and Henry. When they learned of this later, the trustees described this action as a betrayal of their trust. If he had been successful, the trustees would have been left with massive financial commitments and no director, unofficial or official. In their pamphlet defending Gould, the other members of the Scientific Council denied that the astonomer was seeking the Columbia position, but the references are clear in their correspondence. Gould was adept in the delicate steps required in pursuing a position without appearing to do so. He never stated openly that he wanted the position at Columbia, but his letters to his friends reveal his intentions. Gould's denial in this situation is surprisingly analogous to one that would precipitate Gould's disastrous break with Benjamin Peirce in 1867 over the Harvard Observatory directorship. In both cases—Columbia and Harvard—Gould claimed ignorance of the campaign to have his name put forward. In both cases Wolcott Gibbs was active in pressing Gould's interests. Gibbs was one of Gould's oldest and most intimate friends from their days as students in Germany. In 1867 Peirce would not believe Gould's protestations of innocence. In 1858 neither would the Albany trustees.[44]

CHAPTER 6

Conflict Begins: The Peters Problem

In ancient times, when most believed that movements in the heavens had a direct influence on the affairs of mankind, comets were regarded with mingled interest and apprehension. If such beliefs still held, then the crises that befell the Dudley Observatory in 1857 were well foreshadowed. Not only did the arrival of the Great Comet of 1857 arouse fearful interest in Albany, but another comet, barely discernible through a telescope, played its own disturbing part in the events that followed.

Eighteen fifty-seven was a difficult year. The spring began with a devastating flood; the autumn brought a national financial crisis in which the banks of Albany, along with those of the rest of the nation, were forced to suspend payment. The panic subsided in the winter, but reverberations continued through the following year. The flood did not affect the observatory, but the financial crisis had an overwhelming impact. The year was not an easy one for members of the Scientific Council either. Benjamin Peirce was entangled in a distastefully public dispute with Professor C. F. Winslow, who accused Peirce of plagiarizing his work. Peirce's embarrassment was compounded when an amateur mathematician from Pennsylvania, John Warner, accused the Harvard professor of another instance of plagiarism. Peirce maintained a low profile during these attacks, concentrating instead on trying to secure the post of Rumford Professor at Harvard for fellow Lazzaroni, Wolcott Gibbs.[1] Bache and Joseph Henry were preoccupied with their own problems at the Coast Survey and the Smithsonian. Gould remained in Cambridge, working on Coast Survey computations while he carried on his long-distance supervision of the Dudley Observatory renovations.

At the age of thirty-three, Gould's combination of youth and education seemed to offer both the enthusiasm and expertise necessary to bring the observatory into full operation. The endowment grew steadily through January of 1857, and the young astronomer's affiliation with the observatory was still presumed. In lectures to the Young Men's Association and at many of the small evening gatherings to which he was invited when visiting Albany, Gould referred repeatedly to the benefits the observatory derived from its close connection with the Coast Survey. Although unofficial, the association with a federal agency seemed to promise a measure of security to the new observatory.[2]

This special relationship was helpful in encouraging donations, but it was injudicious for any member of the Coast Survey to acknowledge it. Congress expected no such long-term commitment between an observatory and the Coast Survey, whose authorization specifically excluded the use of public funds to establish or support an observatory. With this in mind, Bache and Gould were meticulous in their denials in 1857 that any *official* connection existed between the Coast Survey and the Albany observatory. The only officially recognized relationship was the temporary selection of the Dudley Observatory as the central point in the Hudson River triangulation.

Work on the observatory was still dishearteningly slow. It was difficult for anyone to conjure progress out of the battered appearance of the building. During the spring of 1857 Gould entered another of his recurring periods of depression. He must have mentioned his lassitude to Armsby, for the energetic doctor urged him to fight off his discouraged state of mind. Armsby chided Gould for his loss of spirit, reminding the dejected astronomer that their resources a year ago had included only the building and a small sum for the heliometer. Gould responded by assuring Armsby that he was willing to carry his share of the load but said that he had been criticized in Cambridge for giving far too much time to matters in Albany. Rejecting the intimation that he was "lukewarm," Gould alluded to the mental oppression that weighed upon him, describing a "weariness of spirit" that was breaking him down. This debilitating despondency continued through May. By late spring Gould described his condition as an illness of the brain that left him with "no power to act, to think, or to contemplate. No vigor nor energy nor hopefulness or vital force. Trying to work too hard, something snapped and now I can't work at all."[3]

With Gould in a trough of depression through the spring, the observatory was left to Dr. Peters. By June, Peters was reporting that the building was coming along well and was of the opinion that everything would be ready to receive the instruments in July. Peters was aware that the rapid completion of the Ann Arbor observatory was a constant source of irritation to the trustees. The end result in Albany, he insisted, would be "something of greater importance for science than the hundreds of College-Observatories which were springing up around the country."[4] While directing the workmen, Dr. Peters continued his Coast Survey work with the only operational instrument, a small comet-seeker. In July, Peters made the first discovery for the Dudley Observatory, reporting the original sighting of a telescopic comet, one not yet visible to the naked eye. The discovery of an insignificant comet was not important to the kind of astronomical research Gould planned for the Dudley Observatory. For the enthusiasts in Albany, however, the discovery prefigured a promising future once the institution came into full operation. Unfortunately, this comet's appearance precipitated a series of misunderstandings, the first of many to follow. According to the editor of the *Albany Atlas and Argus,* the comet was the "first fruit" of the Dudley Observatory, but it would end as a bitter apple of discord.[5]

Up through the middle of the nineteenth century, comets were often named for their discoverers, as with Biela's Comet; others were known informally by their size, such as the Great Comet of 1857. The newly-discovered asteroids were receiving mythological names, like Vesta and Hera. Gould thought this practice was unscientific. While studying in Europe, he had expressed his fear that an asteroid discovered by an American might be dubbed for "Franklin, Washington, or some such name." He made his position clear in the scientific community in his 1849 review of John F. W. Herschel's *Outlines of Astronomy* and reiterated his arguments in letters to Airy and other English astronomers. Gould campaigned vigorously to have the burgeoning number of asteroids and comets designated only by the year and sequence of their discovery. Thus, Peters's new discovery should be known as Comet 1857-IV.[6]

Consequently, when Gould's own assistant named the new comet in honor of Thomas Olcott, president of the Mechanics' and Farmers' Bank of Albany, Gould was irritated as well as embarrassed. Dr. Peters was Gould's subordinate in the Coast Survey. Everyone in the scientific community knew of Gould's unofficial attachment to the Dudley Observatory. He wrote Peters at once, saying that although it was a "pretty idea" to honor such an excellent man, it was contrary to all his theoretical principles.[7] Gould's position was unquestionably consistent with his own opinion on proper nomenclature in astronomy. Rapid developments in optical craftsmanship were leading to the discovery of new comets in ever larger numbers; the list of asteroids was expanding steadily. Some organized system of classification putting each into relation with the others would be beneficial, but Gould's suggestions on nomenclature had not yet been adopted. Unquestionably, Peters's decision was unorthodox. Appending a bank president's name to a comet was an unusual step, to put it mildly. Unfortunately, Gould's open dissatisfaction left an impression in Albany that Gould either did not wish to honor Olcott or, even more unfairly to Gould, resented the fact that the discovery had been made by Peters and not by himself. The Olcott Comet became yet another awkward incident in Gould's relations with the trustees. Such perturbations might be overlooked while the bonds of mutual trust still held the Cambridge-Albany relationship in equilibrium. Once the stabilizing force of that trust disappeared, however, old lines of attraction and repulsion would reestablish themselves.

The actions of Dr. Peters in this situation appear as calculated as Gould's seem clumsy. The European astronomer had developed a strong following in Albany in a short period of time. Gould's refusal to leave Cambridge and his lethargic leadership during 1857 contrasted sharply with the energy of the congenial Dr. Peters. Bache had been persuaded to hire Peters for the Coast Survey by Gould, but "the Chief" never felt at ease with the European astronomer. Since hiring him, Bache had been pleased enough with his work to give Peters an increase in salary. Despite their private reservations, both Bache and Peirce had praised the immigrant astronomer highly to the trustees before they sent him to Albany. The trustees were obviously delighted to have him at the observatory. Peters's European training and excellent credentials provided an international panache that fit their vision of the institution's future. Dr. Peters was highly visible, working industriously under orders from the absent Gould. His discovery of the comet, so helpful to the public image of the institution, confirmed the trustees in their positive impressions. By placing the diligent Peters in Albany, the Coast Survey seemed to be fulfilling the first part of the mutual bargain of 1855, and Gould's absence could be reluctantly accepted.

Renovation's plodded on, but the financial crisis spreading across the nation that summer dimmed prospects for raising the full $150,000 stipulated for the endowment by the Astronomer Royal. The endowment had always been the critical criterion, the hinge on which everything would swing for the Scientific Council. In the meantime, as president of the Mechanics' and Farmers' Bank, Thomas Olcott was desperately engaged in steering his own course through the turmoil of the autumn. Despite the financial difficulties that surrounded him, Olcott's behavior reflected the strength of the antebellum standards of stewardship. He remained committed to the observatory and urged that work continue to bring it into operation. As far as the trustees were concerned, even the minimal level at which the observatory had functioned since the inauguration gave encouragement, especially while Dr. Peters remained on hand.

The panic worsened. Although Gould reassured Olcott that he would keep the observatory moving forward, the decision was taken out of his hands in early November of 1857. Fearing possible criticism in financially difficult times, Bache did not want to keep an employee in Albany if the instruments that justified that posting were not yet operational. The meridian circle and transit had arrived but were still in storage. With one eye on Congress, Bache ordered Dr. Peters to return to Cambridge to work under Benjamin Peirce's supervision. The observatory was to be closed indefinitely.[8]

Peters had been the director of a national survey himself and his more menial position, answering to the younger Gould, had probably chafed the European astronomer. He would certainly resent returning to the close supervision of the punctilious Peirce after his independence in Albany. His summary reduction from the more prestigious title of "observer" to that of "computer" was also offensive to a European astronomer who, like Gould, held a doctorate from Göttingen. Perhaps Peters gleaned intimations of Bache's distrust. At any rate his lifelong estrangement from the scientists who made up the Scientific Council began. He wrote his letter of resignation from the Coast Survey. Bache accepted with alacrity.[9]

When Gould heard of Peters's resignation, he was personally offended. He had persuaded Bache to hire Peters and thought the immigrant astronomer should at least be loyal enough not to embarrass his benefactor by resigning precipitously. Gould described Peters's refusal to return to Cambridge as an attempt by a subordinate to dictate the terms of his employment. With all of these feelings of embarrassment, distrust, and intimations of disloyalty, Gould was in no mood to be lenient when Olcott asked if Peters might remain in Albany to set up the meridian circle and transit. As Gould later described his reasons, loyalty to the Coast Survey precluded him "from rewarding disaffection and insubordination; discretion forbade me to trust the man farther; propriety and duty alike withheld me from exhibiting to other assistants an example of faithlessness successful, and the authority of our own official chief insulted."[10]

Although Bache had been delighted with Peters's resignation, the thought that the astronomer might be independently associated with the observatory was another matter completely. Bache was adamant that Peters must have no connection with the Coast Survey, even unofficially through Albany, unless he obeyed the order to move to Cambridge. Although pressed repeatedly by Olcott, the Scientific Council adamantly refused to allow Peters to remain at the observatory.

The scientists expected no opposition to their decision, showing little sensitivity to local feelings. Although the decision to close the observatory brought beseeching letters from Albany, the members of the Scientific Council assumed they were still in control. After all, the trustees had willingly followed all their recommendations before. They were apparently unaware of mounting dissatisfaction with the slow pace of Gould's leadership as well as with his refusal to come to Albany. Bache had based his decision to close the observatory on the needs of the Coast Survey. In Bache's mind the ability of the Coast Survey to influence the direction and development of American scientific institutions was foremost and should not be jeopardized. From their point of view, the trustees found the scientists' refusal to allow Dr. Peters to install instruments already in storage at the observatory perplexing. If the meridian circle and the transit were set up, the observatory could begin to provide services to the railroads and surrounding cities, thus gaining income to sustain itself through financially troubled times. An active observatory would encourage public confidence, avoiding the indignity of another hiatus. As the trustees reasoned, the Coast Survey had an added incentive to keep the observatory moving forward because survey personnel would have the use of two excellent telescopes that would fully justify their presence in Albany. The refusal of the Scientific Council to consider using Peters's services in setting up the instruments seemed inexplicable. As far as the trustees could see, Peters's only offense consisted solely in his unwillingness to leave Albany, in pointed contrast to Gould's refusal to come.

A petition began circulating in Albany. Thirty-four influential citizens signed, including Blandina Dudley, Thurlow Weed, J. V. L. Pruyn, Stephen Van Rensselaer IV, and Thomas Olcott. The signers urged the trustees to appoint Dr. Peters as an observer so that he could set up the instruments and begin observations. The petition was specifically addressed to the trustees as the persons in charge of the institution and was sent on to the scientists. Bache was highly irritated at this continuing insistence that Peters be retained, but began to be a little nervous about his summary dismissal. Had they been too harsh in their treatment of the immigrant astronomer, he wondered? "If the Albanians ask for Peters and he cries *peccati* then what? but he won't and they will not unless he does."[11]

The scientists did not understand the deep-seated opposition to their seemingly logical decision to close the observatory. Bache, in particular, found the Albany resistance to be mystifying. The scientists offered varying theories to each other, wondering if Olcott was perhaps influenced "by some matter that we have not yet got at."[12] Their suspicions of Peters increased as the campaign to keep the observatory open gathered momentum in Albany. Bache complained bitterly of Peters's disloyalty. Gould felt that Peters's attempt to undercut him by securing an independent position at the observatory was particularly traitorous in light of the help Gould extended when the European astronomer arrived in this country. Bache and Peirce worried about the deep hatred Gould held for Peters, as Gould's emotional stability was noticeably precarious during this period. They had viewed the Dudley Observatory as the proper vehicle for his future. Most certainly, they vowed, the observatory should not fall into any other hands, especially not the disloyal hands of Dr. Peters.[13]

As the growing strength of Peters's support emerged, the

members of the Scientific Council remained perplexed. The endowment was still an all-or-nothing proposition as far as they were concerned. Gould assumed that everyone knew he would not tie himself to anything other than a first-class observatory. The Scientific Council's definition of a first-class institution was premised on the sum specified for the endowment by Airy. Without that full amount, they believed that the Dudley Observatory should remain closed. The difference in perspective is clear. Bache, Gould, and Peirce seriously underestimated the psychological effect on the community of their decision to close the observatory. The trustees had personal commitments and sensitivities that transcended dollars tied to the observatory. They had lent their names to the institution and, as trustees drawn from a group in whom the ethic of stewardship waxed strong, believed they had a duty to see it through this crisis. Their responses to Bache's decision, especially with Olcott, have to be seen in relation to antebellum standards of fiduciary responsibility, standards that involved a personal obligation to see the task through once a commitment had been made. The observatory had been launched once in 1852, languished, and launched a second time with high-flown rhetoric and full publicity in 1855. A second suspension would be resisted by those who had resisted the first so stubbornly. Much of the support for Peters arose simply from his continued presence as a practicing astronomer in an observatory that was actually functioning, however minimally. As the trustees saw it, Peters's presence maintained rather than endangered the future prospects of the institution. They couldn't understand why the scientists couldn't see that also.

As Olcott wrote increasingly firmer letters to Bache, the riddle grew more frustrating to the scientists. If the trustees were capable of such bullheaded resistance to the authority of the Scientific Council, Bache confessed to reservations about sending Gould to Albany. Gould had declined severely, overcome by anxieties. He wrote the Astronomer Royal that the difficulties with Peters and the observatory were "sufficient to explain a severe cerebral attack from which time alone can entirely bring relief." With Gould's condition in mind, Bache seriously weighed cutting loose completely from the Dudley Observatory. Yet he was reluctant to give up an establishment that held promise of becoming a fine research institution. The potential had to be balanced against the irritation. He also had to consider what might happen if the observatory, superbly equipped, should pass into someone else's control. "Why it is a great burglary, darling Function [Peirce]. A national observatory with Peters as chief. Can't be, Function, dear. . . . We will not die without a sign!"[14]

Bache concluded that Peters had somehow poisoned the minds of the trustees. He urged Peirce to go at once to Albany for firsthand information on both the extent of Peters's support and the tractability of the trustees. Once he arrived, Peirce laid most of the blame for the problem squarely on Peters. Bache grew more concerned. With a loyal assistant in Albany or with the observatory closed, Gould could have deferred leaving Cambridge for several years, but Gould's presence in Albany was indispensable if the trustees insisted on keeping the observatory open. It was now clear that if Gould were not there, Peters would be. Despite Gould's fragile condition, he would have to go to Albany; but the state of Gould's health made the retention of Dr. Peters, in any capacity, impossible. Bache feared that the strain on Gould might be suicidal. Peters should be allowed to "eat humble pie" and then return ignominiously to Cambridge when he was unable to find employment elsewhere. Bache castigated James Hall for his continuing support for Peters, observing that Hall should show more gratitude to those who had given him favors in the past.[15]

Gould traveled to Albany to meet with Olcott. In the meantime the scientists conferred through the mails. Without doubt, the members of the Scientific Council had valid grounds for their hesitation with regard to the Dudley Observatory. Despite the impressive early donations, the financial crisis of 1857 had dried up the stream of contributions. Although the observatory had an endowment of substantial size for an American institution, it fell short of the figure set by the Astronomer Royal. Because of the extensive expenditures on instruments and remodeling, two years' dividends would be required to restore the endowment to its original condition. Until the observatory could provide services to the community, the only other available source of funding was a two thousand dollar appropriation from the New York legislature for the observatory's longitude determination.

When Gould arrived in Albany in mid-December, Olcott suggested that part of the legislative appropriation be used to pay Peters's salary, so that he could set up the instruments already at the observatory. Gould refused. Olcott then proposed that Gould move to Albany and take Peters on as an assistant. Gould again declined, warning Olcott that if the European astronomer remained at the observatory, no further aid could be expected from the Coast Survey, now or in the future. Olcott then asked if two Coast Survey assistants might be assigned to the observatory immediately. If Peters remained, Gould responded, no Coast Survey personnel would serve with "one who had been guilty of gross disobedience of orders." Olcott replied to this thinly veiled threat with one of his own. Rather than have Peters leave the observatory, Olcott would sever connections with the Coast Survey altogether. Unless Gould was willing to come to Albany, Olcott could see no way to satisfy the community, which was clamoring for Peters.[16]

Benjamin Peirce was disgusted. He thought Gould had mishandled the entire meeting by making such open refusals. Despite the scientists' irritation, Gould's meeting with Olcott convinced Bache and Peirce that the observatory would remain in operation either with or without the approval of the Scientific Council. Their alternatives were clear. The observatory could proceed under the direction of either Gould or the hated Dr. Peters. The answer was obvious. Gould must move to Albany; Bache would send several Coast Survey assistants with him. Irritated beyond measure, Bache confessed that his temper almost surpassed his judgment at this point. He reiterated his strong desire to "shake off the dust of the affair," but the vision of the possible future awaiting the observatory held him bound to Albany.[17]

Bache and Peirce were aware that their presently undefined status as scientific advisors did not give them firm control of the observatory, now or in the future; instead, the authority of the Scientific Council clearly depended upon amicable relations with the trustees. The frustrating events of the past few weeks pointed out the problems with this informal arrangement. To secure control, Gould must now assume the title of director. Gould could then formally call upon the Scientific Council for advice "and we will direct the affair . . . meanwhile hedging in our domain so that it may all be snug."[18] Bache was convinced that once Gould was in residence in Albany, he would be able to bring the trustees into line. Despite the bumpy ride of the autumn, the Scientific Council would still be in the saddle.

Pursuing threads of information gathered during Gould's mid-December meeting with Olcott, Bache deduced that Armsby and Olcott were reluctant to convene the full board of trustees. He surmised that the erosion of the endowment might be embarrassing if made public. Because the full power of the board could not be called upon, Bache reasoned that the trustees would be unable to interfere officially with Gould's actions as director. "If this is not a check mate than [*sic*] never was one."[19] Gould hurried off to Albany, however reluctantly. The tangle over Peters seemed to have been straightened out, although Bache was now aware that there had been more feelings aroused against the Scientific Council than he had estimated. Shaking his head over such a poor reward for his efforts in behalf of the observatory in particular and American science in general, he returned his attention to his own thorny problems with the Coast Survey appropriation and the House Ways and Means Committee.

Gould was left with the unenviable task of working everything out in Albany. Peters remained ensconced in the observatory and showed no signs of leaving. Wherever Gould went, Dr. Peters's friends continued to press for his appointment. Finding face-to-face refusals increasingly difficult, Gould fell back on the authority of the Scientific Council, saying that he would abide by their advice on the question. The council's answer to such a query was already obvious, but he preserved the form of appealing to a supposedly impartial tribunal. Gould observed that his "ungenerous crushing of the poor foreign martyred laurel crowned man" had made his own course in the city an uphill battle.[20]

Gould spent Christmas in Cambridge. One of his assistants, James Toomer, remained in Albany. Gould believed that his meetings with Olcott had resulted in a mutually acceptable agreement about the observatory. He planned to return in mid-February as head of a Coast Survey longitude team based at the Dudley Observatory. Apparently, there was no public knowledge of these arrangements. Strong feelings in favor of Peters's appointment continued to grow during the holiday season. Some implied that the disgraceful manner of Peters's dismissal would interfere with further donations. Because no permanent arrangement had been announced publicly, the observatory situation appeared to be unresolved.

It is certainly possible that Gould did not hide his reluctance to move to Albany during December, diminishing Olcott's faith that the observatory would have an astronomer in residence. Perhaps Olcott liked Dr. Peters too much to let him go. The Coast Survey assistant left in Albany by Gould was a young man of nineteen or so, who must have suffered by comparison with the mature European astronomer. Matters moved quickly in the first few weeks of the new year. Peters refused to leave his quarters at the Dudley Observatory, forcing Toomer to reside in a hotel at Coast Survey expense. After a week or so, Toomer sent the alarming news to Cambridge that not only was Peters still in residence, but he had also unpacked his books and ordered a full winter's supply of coal. If Gould was to displace the dastardly Peters and install Toomer in the observatory, he had no choice but to board the train to Albany.

On his arrival Gould promptly penned a formal note to Peters, asking when he planned to leave the observatory. Peters responded with the devastating news that he had been asked by the trustees to remain in the building. He stated flatly that he would not leave the observatory unless requested by the trustees to do so. Peters was no longer employed by the Coast

Survey, and Gould no longer had any authority over him. Because Gould was not officially the director of the Dudley Observatory, the trustees' request that Peters remain in the building could not be countermanded.[21]

Gould knew he had to tread carefully. Petitions urging the trustees to appoint Peters were still circulating, already signed by Stephen Van Rensselaer IV, Thurlow Weed, and J. V. L. Pruyn. Pruyn noted in his journal that he saw no difficulty in such an appointment. James Hall was still actively supporting Peters also, and his word carried weight in Albany. In a convoluted argument, Hall maintained that although the Scientific Council would not consent in advance to the appointment of Dr. Peters, no trouble would come of it if the board decided to go ahead. Hall was also rumored to have criticized Gould for being so slow in bringing the observatory into operating condition.[22]

The board of trustees met on 9 January 1858. During the heated session, several letters from Peirce and Bache denouncing Dr. Peters were read aloud, prompting sharp comments from many trustees who vividly recalled the scientists' glowing endorsements of the astronomer one year earlier. Gould's personality also came under discussion in the meeting, especially his temper. Gould's sharp-tongued rejections of the attempts of several unnamed trustees to involve themselves in the remodeling received open criticism for the first time. It is clear that the trustees had been stung by Gould's arrogance and frustrated by his absence. The year in which Gould insisted on remaining in Cambridge had exacted a price. As an indication of their own reaction to his attitudes during the past year, the board expressed its gratitude to Dr. Armsby for his unremitting efforts in behalf of the observatory. The members of the Scientific Council were then formally appointed as official advisers to the observatory.[23]

After the meeting Gould found Olcott and Robert Pruyn together. The two trustees again urged Gould to take Dr. Peters on as his assistant. Again he refused, saying that the Scientific Council as a whole had termed further employment of Dr. Peters to be "inexpedient."[24] He wrote Peirce that he wished heartily that "that snake [Peters] were only away from the observatory," because Armsby would then be powerless. As long as Dr. Peters remained, Gould felt that the astronomer and Armsby would act together against him. Although anxious to get Peters out, Gould did not wish to appear vindictive. Begrudgingly, he agreed that Peters might remain another week or so. He tried hard to "let Mr. Olcott down easy," but felt there was nothing more he could do to oblige him. Gould believed that Olcott had taken this final refusal to hire Peters "kindly and well," so he left for Washington.[25]

On his way home to Cambridge from Washington, Gould received the devastating news that Dr. Peters had been appointed as an "Observer" by the trustees after he left Albany. In the official announcement, the trustees expressed their appreciation for the cooperation and advice of the Scientific Council, but reaffirmed their "undivided and entire control" of the observatory. They emphasized that donations had been used to purchase costly instruments that were now in storage. Future donations would be solicited on the basis of the institution's ability to contribute to the advancement of astronomy and to supply public services. Referring to their own strong sense of responsibility as trustees and to the wishes of many of the donors who wished to bring the observatory into operating condition as quickly as possible, the board appointed Peters to a permanent position.[26]

Was this an attempt by the trustees to sever relations with the Scientific Council? Although the appointment would not smooth relations with the scientists, it is doubtful that the trustees intended to cut off the Scientific Council completely. Olcott and Armsby knew that success for the observatory depended upon amicable relations with the influential members of the Scientific Council. But they were even more adamant that the observatory be brought into operating condition at a faster pace than Gould had set. From the trustees' perspective, the scientists did not seem either to act in the best interests of the observatory or to feel the same urgency. Their announcement that the institution would be closed, without any prior discussions with the trustees on the subject, seemed arrogant and arbitrary. Their passionate refusal to tolerate Dr. Peters's presence also lessened the scientists' credibility. The trustees knew the value of the connection with the Scientific Council; but they were also keeping their first goal in mind—establishing a working observatory. Peters was a competent, well-educated astronomer who wanted to remain in Albany. His appointment by the trustees was an insistent affirmation that the observatory would go ahead.[27]

Gould was in despair. Bache and Joseph Henry were furious, placing major culpability on the meddling interference of James Hall, Armsby, and the nefarious Peters. The actions of the board were attributed to Armsby's influence and Olcott's acquiescence. The question now facing the scientists was how to turn public opinion away from Dr. Peters. Gould's friend, John Gavit, hurried to Washington from Albany to enlist the influence of Erastus Corning, congressman from Albany and president of the New York Central Railroad. Corning was a close friend of Joseph Henry's, having served for years as the Henry family banker. He was a powerful New York politician as well as a business and political rival of Thomas Olcott.[28] Bache and Peirce decided that Joseph Henry, so well respected in Albany, must persuade James Hall to help undo his unwitting damage. Likening themselves to the Light Brigade, the scientists prepared to advance up the Hudson. In order to preserve their future control of the Dudley Observatory, they were willing to yield on all matters at this point—except for Dr. Peters. Peters's presence could not be tolerated. Treachery must not be allowed to succeed, they wrote. Meanwhile, Gould would remain in Cambridge to avoid the appearance of working in his own interests. The scientists believed that this maintained "a certain appearance of delicacy" in contrast with the crass self-interest they attributed to Peters. Convinced they were setting out to do battle in the full armour of righteousness, Henry, Peirce, and Bache converged in full panoply on Albany, ready to "charge and 'die or do.'"[29]

Despite the adulation offered to Joseph Henry on his arrival, it was obvious that the situation in Albany was more serious than the scientists expected. Putting it mildly, they found that "a state of unpleasant feelings existed in regard to Dr. Gould." Although the scientists thought that Olcott had broken his agreement with Gould, they did not wish to alienate the powerful bank president. Instead, they approached other members of the board. Gould suggested that Judge Parker and Ezra Prentice were two trustees whose opinions carried weight in Albany and could be "put right." Judge Harris, however, was "too thoroughly Peterified" to convert."[30] In many private meetings, the scientists tried to persuade individual members of the board to dismiss Peters and appoint Gould officially as director. The trustees insisted that they could no longer accept the slow pace of the past eighteen months after so large an amount had been invested in the institution. Too many promises had been made to the public. Those who "wished to hold positions of honor in the Observatory, could make up their minds to put it in operation, with the means already secured" or the relationship must terminate. In response the members of the Scientific Council denounced Peters vehemently and explained Gould's inactivity by referring to his Coast Survey responsibilities.[31]

The board convened officially on 16 January. Gould arrived in Albany that morning. Each of the scientists addressed the meeting. Henry spoke "in a strain of highly impassioned eloquence"; Peirce spoke "coolly and dispassionately" of the need for harmony. Bache struck hard at Peters. Gould then took the floor. In retrospect Gould's remarks at this meeting serve as a watershed in his never-easy relations with the trustees. He reviewed the past two years, then proceeded to describe the indignities he had suffered, "doing this, however as kindly as was possible under the circumstances, and giving but a small part of the whole truth." Most of the trustees listening believed Gould's speech to be a pointed criticism of Dr. Armsby. In addition to this unexpected personal attack on a trustee, Gould went on to say that he had suffered continual annoyances whenever he came to Albany. His plans had been subject to constant interference, making his position in Albany "a bed of thorns." When Gould described his experiences in Albany as a bed of thorns, he was probably expressing his feelings in as restrained a manner as he could assume. Everyone on the board, however, could recall the many parties, gatherings, and adulatory assemblies that had greeted the scientists on every visit. Gould had never made his enormous resentment of Dr. Armsby's efforts public. Most of the trustees assumed Armsby and Gould were still "warm friends," recalling the fulsome praise he had heaped on the physician at the dedication ceremonies. Most believed it was unnecessarily humiliating for Armsby to have these embarrassing and unexpected accusations uttered before the entire board.

The trustees had spent enormous sums of money under Gould's direction. They had deferred to every suggestion handed down from Cambridge. Gould's declining interest in the observatory during 1857 and his seemingly arbitrary refusal to allow Peters to be associated with the institution had undermined their trust. Now, as Gould continued speaking, he destroyed that trust completely with a stunning disclaimer. The astronomer made the astounding declaration "that he *was not responsible for anything which had been done at the Observatory—that the improvements and alterations and erections—I think he broadly included nearly all expenditures*—had been not only against his wishes but in opposition to his expressed objectives."[32]

Gould had ordered all the observatory's instruments to his own particular specifications, had drafted plans for a dwelling house, as well as plans for an extensive reconstruction of the original observatory building. The astronomer also had plans drawn for four or five additional buildings to be constructed in the future, with the costs estimated at between $100,000 and $200,000. Olcott was especially sensitive to the fiscal condition of the observatory and the extent to which the endowment had been impaired for renovations and equipment. If Gould disavowed responsibility for these decisions, the trustees would shoulder the complete burden for these expenditures and commitments. With his disclaimer of responsibility, Gould violated the basic standards by which businessmen like Olcott operated.

In the volatile financial climate of antebellum America, morality was closely tied to fiscal responsibility. Fiduciary responsibility reflected personal integrity and provided the basis for trust.

This was especially true in the 1840s and 1850s when speculation had been rampant and state legislatures toyed with repudiation of debt. Anyone in business during that time was keenly aware of the daily problems he faced in a chaotic system of state bank currencies with widely fluctuating values. In the eyes of men like Thomas Olcott, the economic stability of the nation depended on the personal acceptance of obligations. A man's word was a pledge. Dishonoring such a pledge, after others had placed their trust and financial security in a man's hands, was a moral injustice. Olcott viewed the repudiation of personal obligation as immoral and unchristian, if not a direct threat to the economic security of the nation. A strong supporter of the temperance movement and a growing power in the Republican party, the bank president was clearly part of that conservative group of reformers who hoped to restore the nation to the path of righteousness. Gould's disclaimer raised serious questions in Olcott's mind about the astronomer's integrity.[33]

Olcott had more immediate personal concerns also. As a result of Gould's declaration, Thomas Olcott's financial reputation became a large part of the stakes in the public's perception of the observatory's difficulties. Bache had earlier recognized the vulnerability of the trustees on the money question, but the point was driven home to Olcott by Gould's remarks. A man recognized as the greatest banker in New York would not take the risk of public embarrassment on money matters lightly. The prospect of public inquiry into the observatory's present condition, having spent beyond its means, was not only potentially embarrassing but also threatened a reputation for financial probity that had taken years to build.

The discussion that followed Gould's speech was heated, with the board dividing into a majority of nine, which followed Olcott's lead, and a minority of four, which included business associates of Erastus Corning, supporting the Scientific Council. The trustees adjourned without action. Olcott remained stubborn in his support for Peters, but a supposedly "grand reconciliation" between Gould and Armsby was arranged at a party that evening at General Robert Pruyn's home. Gould and Armsby shook hands in an upstairs room while Olcott and Joseph Henry conferred in the downstairs library. The following day Robert Pruyn, Bache, and Joseph Henry met again with Olcott. This time the bank president "caved in." He agreed to rescind Peters's appointment, if it could be done gently. By donating the amount of Peters's salary, Olcott arranged to send the astronomer to Hamilton College's new observatory. The trustees later stated that they dismissed Peters "for the sake of harmony" and out of their regard for the Scientific Council. The scientists denied that Peters had been sacrificed to Gould. Instead, they maintained that Peters was not tolerable because of his incompetence and his moral failings. Bache and Henry visited Armsby to tell him that Peters had been dismissed and Gould restored. According to J. V. L. Pruyn, this second dismissal of Dr. Peters cemented Armsby's antipathy toward Gould. The "grand reconciliation" had failed. Armsby was bitter. Gould returned those feelings in full measure, confiding to Peirce that the doctor would "leave no stone unturned[,] no tricking fraud unperpetuated to trip up our heels yet."[34]

The trustees reconvened on 19 January. Olcott stated that if Peters had to leave, Gould must take up residence at the observatory immediately. He also stipulated that no money could be spent from observatory accounts for the next two years, except for five hundred dollars set aside for the installation of the instruments already purchased. Under the agreement reached at this January meeting, the Coast Survey would maintain several assistants at the observatory who would mount the instruments under Gould's supervision when not occupied with their regular work. The Scientific Council was to assume full control of the scientific work of the institution. But the boundaries of power were never spelled out. The only real understanding on which everyone agreed was that Peters would leave and Gould would come. Although they had ousted Peters and kept control of the observatory, Peirce cautioned Bache that their position had to be better guarded: "*Real power must be secured,* if it is possible. For this purpose, bonds should be given for the property entrusted to the Council or to the Coast Survey, or to the Director, or to all."[35]

The January arrangement was one of those rare instances in which both sides emerged with a sense of victory. The trustees had turned back the attempt to close the institution and, at the same time, managed to secure an increase in the observatory's professional staff at no cost. The Scientific Council could equally claim victory and did. As Bache exulted to Peirce, "Yes the plot *was* in vain, and so finish all plots against the Lazzaroni."[36] The dastardly Dr. Peters was banished. The Dudley Observatory was Gould's to steer to its future. The Scientific Council later claimed that this January agreement was a binding contract that could not be annulled except by consent of both sides. Yet Gould was himself uncertain about how long the Albany situation would continue to suit his own plans. The members of the Scientific Council still felt free to cut themselves loose from the Dudley Observatory; but they would not allow the trustees to sever the connection.

The compact that purchased peace in January of 1858 was packed full of problems. Essential questions remained unresolved: Who possessed ultimate control of the Dudley Observatory? What were the participants' expectations about authority and responsibility? These issues were never confronted directly in this January agreement. Obviously, the resolution of the January conflict meant different things to the participants. The trustees opposed the closing of the observatory. The scientists opposed Dr. Peters's role in the institution. Both sides claimed victory. But had control of the observatory passed from the trustees to the scientists? Or, even more importantly, had control of the observatory passed to the Coast Survey, an agency of the federal government? Public assumptions of "who holds power" would be important as the controversy over the observatory developed. When professionals attempted to expand their authority in the nineteenth century, preexisting power structures were crucial determinants. In nineteenth-century institutions dependent on voluntarism—colleges, hospitals, libraries—it was generally assumed that final authority rested ultimately with the trustees. In their appointment of Dr. Peters, the trustees specifically referred to the power vested in them by the state legislature to control the institution. Could such power pass to the Scientific Council which had itself been appointed by the trustees and not be explicitly discussed?

What was the result of the controversy over Dr. Peters? The observatory remained open and increased its resident staff. A specific commitment was made by the scientists to install the instruments as soon as possible. Gould, now reluctantly a resident in Albany, was no longer trusted by Olcott. Dr. Armsby hated Gould intensely. His feelings were reciprocated with a vengeance by Gould. In retrospect the conflict over Dr. Peters was an obvious turning point in relationships among those associated with the Dudley Observatory. As George Bond noted, "The Scientific Council made a sad mistake in quarreling with him [Peters]. He had a capacity for the controversy which they never reckoned upon."[37] The lines of the later conflict were drawn.

1. Dr. James H. Armsby, from the Collections of the Albany Institute of History and Art

2. Alexander Dallas Bache, courtesy of the Smithsonian Institution

3. Ormsby Macknight Mitchel, reproduced from the Collections of the Cincinnati Historical Society

4. The Dudley Observatory, courtesy of the Trustees of the Dudley Observatory

5. Benjamin Apthorp Gould, Jr., courtesy of the Harvard University Archives

6. Benjamin Peirce, courtesy of the Harvard University Archives

7. "Dudley Observatory Dedication," by T. H. Matteson, from the Collection of the Albany Institute of History and Art

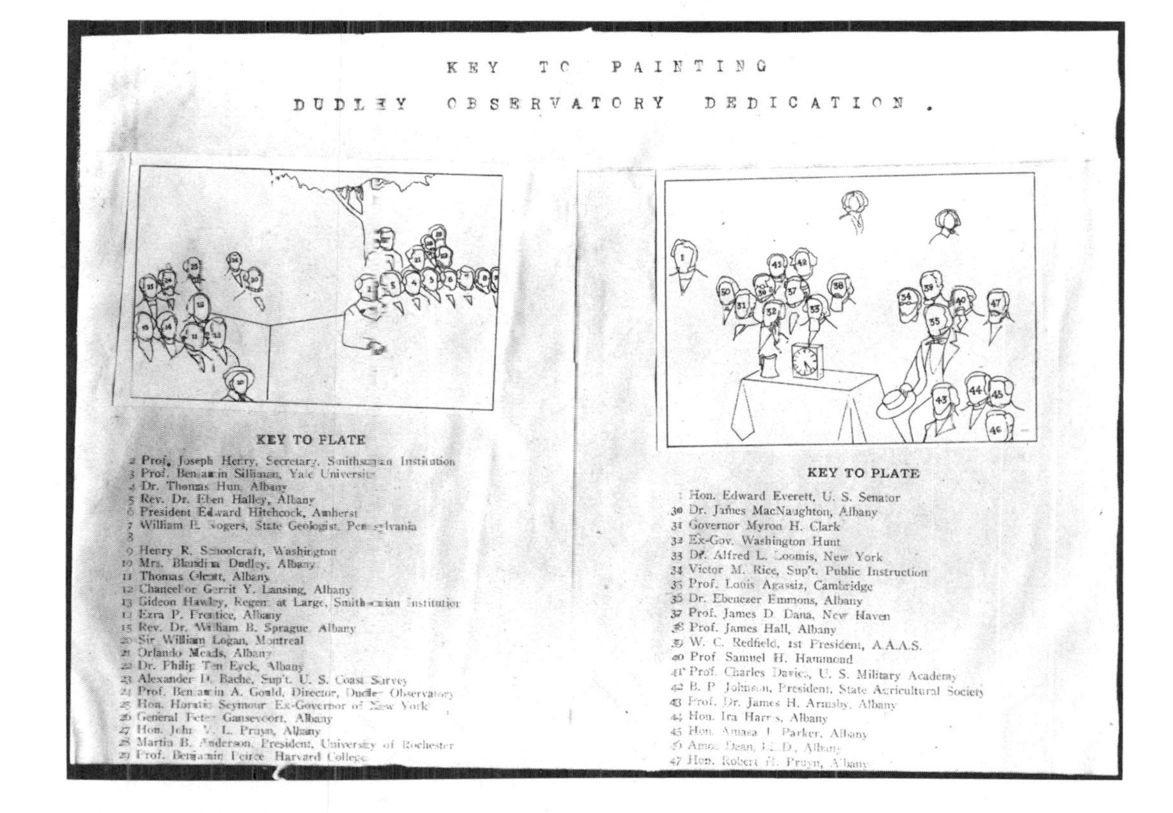

KEY TO PAINTING

DUDLEY OBSERVATORY DEDICATION.

KEY TO PLATE

2 Prof. Joseph Henry, Secretary, Smithsonian Institution
3 Prof. Benjamin Silliman, Yale University
4 Dr. Thomas Hun, Albany
5 Rev. Dr. Eben Halley, Albany
6 President Edward Hitchcock, Amherst
7 William B. Rogers, State Geologist, Pennsylvania
8
9 Henry R. Schoolcraft, Washington
10 Mrs. Blandina Dudley, Albany
11 Thomas Olcott, Albany
12 Chancellor Gerrit Y. Lansing, Albany
13 Gideon Hawley, Regent at Large, Smithsonian Institution
14 Ezra P. Prentice, Albany
15 Rev. Dr. William B. Sprague, Albany
20 Sir William Logan, Montreal
21 Orlando Meads, Albany
22 Dr. Philip Ten Eyck, Albany
23 Alexander D. Bache, Sup't. U. S. Coast Survey
24 Prof. Benjamin A. Gould, Director, Dudley Observatory
25 Hon. Horatio Seymour, Ex-Governor of New York
26 General Peter Gansevoort, Albany
27 Hon. John V. L. Pruyn, Albany
28 Martin B. Anderson, President, University of Rochester
29 Prof. Benjamin Peirce, Harvard College

KEY TO PLATE

1 Hon. Edward Everett, U. S. Senator
30 Dr. James MacNaughton, Albany
31 Governor Myron H. Clark
32 Ex-Gov. Washington Hunt
33 Dr. Alfred L. Loomis, New York
34 Victor M. Rice, Sup't. Public Instruction
35 Prof. Louis Agassiz, Cambridge
36 Dr. Ebenezer Emmons, Albany
37 Prof. James D. Dana, New Haven
38 Prof. James Hall, Albany
39 W. C. Redfield, 1st President, A.A.A.S.
40 Prof. Samuel H. Hammond
41 Prof. Charles Davies, U. S. Military Academy
42 B. P. Johnson, President, State Agricultural Society
43 Prof. Dr. James H. Armsby, Albany
44 Hon. Ira Harris, Albany
45 Hon. Amasa J. Parker, Albany
46 Amos Dean, LL.D., Albany
47 Hon. Robert H. Pruyn, Albany

8. Key to the Dudley Observatory Painting, from the Frick Art Reference Library

9. Thomas Worth Olcott, from the Collection of the Albany Institute of History and Art

10. Joseph Henry, courtesy of the Smithsonian Institution

CHAPTER 7

Money, Misunderstandings, and Mistrust

Lack of money can strain the best of relationships. The marriage of the director and trustees of the Dudley Observatory already had difficulties that compounded as cash grew scarce. Dr. Armsby remained resentful about Peters's humiliating dismissal and Gould's insulting attack during the January meeting. Dr. Alden March, Armsby's brother-in-law and a trustee as well, was equally outraged. John Wilder, another member of the board, was furious about the scientists' insistence that Peters depart. The January conflict over Dr. Peters had precipitated vague dissatisfactions. Distrust now crystallized in quarrels over expenditures.

James Hall was aware that the animosities aroused in the disagreement over Peters remained strong but believed that hard feelings, although annoying, presented no real obstacle to the future of the observatory. He reassured Gould that Armsby and Olcott remained dedicated to the success of the Dudley Observatory and would give the astronomer whatever assistance he needed for the good of the institution. Hall underestimated the impact of the January conflict over Peters and, as a result, would be unprepared for the later explosions of hostility.[1] In retrospect it is obvious that the conflict over Dr. Peters changed one factor essential to the future of the Dudley Observatory. Olcott's belief that the astronomer had repudiated his responsibility for money spent in equipping and remodeling the observatory to meet Gould's own specifications severely altered the bank president's attitude. Gould's unexpected attack on Dr. Armsby in the same meeting was also critical. The two most influential trustees no longer trusted the astronomer from Cambridge.

The greatest indictment against Gould stood on the highest ground in Albany. Despite the enormous sums and effort expended, the observatory remained inoperative. The Ann Arbor Observatory, established at approximately the same time as the Albany institution, was already fully functional. Thousands of dollars had been spent by the trustees of the Dudley Observatory since the 1855 agreement with the Scientific Council to bring Gould's image of a "first-class" institution into reality. Given the financial crisis enveloping the nation, the trustees no longer felt easy about dipping into principal, especially after Gould's disclaimer. As the trustees reasoned, if the observatory could be brought into service immediately, it could earn enough income to carry itself through the current crisis and restore public confidence in its future.

During the January trustees' meeting, Olcott had been explicit in his insistence that no additional incursions into capital could be made. Only the two thousand dollars appropriated by the legislature for the observatory's longitude determination was available as a source of immediate funds. Once that survey was completed, any money remaining could then be used to set up the instruments already in storage. In the trustees' minds, the way was open to launch the Dudley Observatory on its official career as a functioning institution through judicious use of the state appropriation.

Those hopes were dashed by a letter from Gould. The astronomer presented a list of expenses totaling $1,500, which he insisted were the minimum requirements necessary to bring the observatory into working condition. The longitude survey would cost about $600, but the trustees now learned from Gould that he did not intend to use the remaining $1,400 to set up the meridian circle and transit so that the observatory could sell time to railroads and nearby cities. Gould instead proposed that the remaining money be spent on additional equipment, which would lie dormant until a later time. To add to their consternation, the trustees were formally advised by Gould a week later that the base of the observatory dome was defective. The astronomer's news dropped like a bombshell; the dome would have to be completely replaced. The trustees were now faced with the demolition of an observatory dome that had never been used. Not only had the walls of the original building come down, but the dome would also fall. The very building seemed to be regressing. The trustees were, in effect, building the observatory twice without having ever seen it work.[2]

What had happened? The dome design, copied from the Russian Imperial Observatory at Pulkova, provided for a system of moveable wooden shutters so lightweight that they could be easily pulled by hand. A similar system was in operation at the Harvard Observatory. At the time of construction, the shutters were not installed. The opening had been covered with tin sheathing. In his modifications of the building, Gould discarded the idea of wooden shutters, which could have been constructed and installed for $300. Instead, he directed the architect to devise a system of flexible iron shutters as well as the machinery capable of moving them automatically. The Scientific Council later defended Gould's innovation as an elegant contribution to observatory architecture. Unfortunately, when the workmen attempted to install the iron shutters in late January of 1858, the dome proved incapable of supporting the weight. During installation the dome aperture widened by several inches. Because of the structural damage incurred, the dome was now incapable of bearing the wooden shutters, much less the iron ones. Rather than discard the iron shutters and machinery, which had already cost about $5,000, Gould insisted that the entire observatory dome be replaced, saying that he had always been convinced that the original dome would be a serious obstacle to the use of any instrument. Gould argued that $1,800 could not be better employed than by replacing the entire dome. In the meantime the astronomer ordered all work at the observatory suspended until the dome was rebuilt. He refused to set up any instruments because the dust stirred up in the dome replacement would damage the telescopes.[3]

Construction of the original dome had cost about $2,000. Estimates for replacing the dome with one meeting Gould's requirements set the cost at about $3,000, almost double the $1,800 he surmised. This dismal news came at a time when there were no funds available for any capital improvements, only the longitude appropriation that the trustees had hoped would be used to bring the observatory into service. The meridian circle and transit still rested in their packing crates at the observatory. The trustees saw this as another instance in which Gould's arrogance and "particularity" worked to the detriment of the observatory. They later charged Gould with trying to boost his own prestige in constructing "something beyond the magnificence of the Royal Institutions of Europe."[4] Some of the trustees were also of the opinion that he was using the dome as an excuse to postpone the installation of the telescopes already on hand. If such a suspicion did exist, it reveals the extent of distrust that had grown among the trustees with regard to their director.

The foundation stones that Gould had ordered for the observatory's instruments were the largest that have ever been ordered, before or since, for observatory telescopes; they were clearly beyond the necessary requirements set by other astronomers. Because of Gould's absence in Cambridge the previous year, the trustees had ended up purchasing two sets of the mammoth stones. Now his shutter system had destroyed the dome. Whatever Gould's motives, both the stones and the iron shutters are appropriate symbols of Gould's approach to the Dudley Observatory. Gould's designs were on the "cutting edge" of observatory architecture of the 1850s. The question has to be asked, however, whether it was practical to insist that every part of the Dudley Observatory be a state-of-the-art exhibit, a testimony to Gould's own professionalism. There are two perspectives on this question. Obviously, from Gould's point of view, the institution had to be excellent in all respects—from the foundation stones to the dome—to become the preeminent American observatory he intended. As far as the trustees were concerned, common sense demanded a more functional and flexible approach during a time of obvious financial difficulties. Let the observatory establish its reputation through its instruments and the work they produced rather than the painstaking perfection in every detail of its construction Gould insisted upon. Public support gained in the interim, they reasoned, would underwrite its higher and long-term purposes. Psychologically, it is also obvious that bringing the observatory into service as soon as possible would give the trustees some sense of momentum, some sense of success in a project that started years earlier. Instead of a sense of progress, the trustees were now faced with a list of necessary expenditures far in excess of their means, plus the disastrous additional cost of a new dome brought about by Gould's ill-fated iron shutter "brainstorm," as they described it. To complete their frustration, the astronomer had ordered all work at the observatory stopped.

The trustees reached a decision quickly. The dome would have to wait and, secondly, work must be resumed at the observatory. They were adamant that the observatory continue to move forward. The telescopes in storage were, after all, supposed to be mounted in the wings, not under the central dome. The new dome would be needed only when the heliometer arrived. Work on that instrument would take several years, and Spencer had not even started. If the meridian circle and transit could be set up to generate some income from railroads and nearby cities in the meantime, no further inroads on the endowment would be made and the observatory would be in a better position to attract public support.[5]

Gould later connected this refusal to approve the new dome with the animosities arising from Peters's dismissal. He also attributed some of the disagreement to private motives. According to Gould, one of the trustees, General Robert Pruyn—a political power as well as a wealthy iron merchant—spoke to him privately, urging that the dome construction be promised to Pruyn's own foundries. Gould implied that Pruyn promised that a dome meeting the astronomer's precise standards would be approved by the trustees if Gould would accept Peters as his assistant. Gould maintained that his refusal to accept Peters or to allow Pruyn to make a profit on the dome was the basis for the rejection of the dome construction project by the trustees and Pruyn's later hostility.[6]

Meanwhile, the Scientific Council and fellow Lazzaroni met in Philadelphia at the home of John Fries Frazer. After dinner Joseph Henry, Peirce, Bache, Gould, Louis Agassiz, and Frazer put their heads together to consider the Albany observatory. The January agreement with the trustees was closely examined by the assembled scientists, who decided that the trustees' explicit reference to Bache as superintendent of the Coast Survey must be deleted. The Dudley Observatory should have no direct claim on the Coast Survey. This explicit denial of any official connection between the survey and the observatory would serve as a pointed contrast to future claims by the scientists. In February of 1858, however, Bache was wary of running the risk of congressional criticism by acknowledging a direct association with an observatory. His friends agreed. The Dudley Observatory was to be kept within their influence, but it must not be allowed to embarrass the Coast Survey.[7]

During the spring of 1858, Bache and Gould continued to deny that the Coast Survey had any official connection with the Dudley Observatory. The only official relationship, they insisted, was the longitude determination. Yet the attempt to divide Bache's association into two distinct and unconnected parts—one as superintendent of the Coast Survey and the other as member of the Scientific Council of the Dudley Observatory—was ineffective. The Coast Survey connection had been repeatedly described as vital to the success of the observatory. An official Coast Survey field party was clearly in residence, and the relationship seemed obvious to everyone despite the meticulous denials. To add to the scientist's irritations, the infamous Dr. Peters was said to be spreading the story over Albany that the Coast Survey was running the Dudley Observatory.[8]

Curiously, the members of the Scientific Council seem to have honestly believed that complete authority over the observatory had passed into their hands through the January arrangement with the trustees. Such a shift in authority would certainly have deserved some official statement of intent from trustees legally entrusted with responsibility for the institution by the state. Despite the absence of such a statement, or even the intimation of one, Bache, Peirce, Henry, and Gould later argued that the January agreement was a legal contract giving the observatory into their hands for two full years, "a contract which morally, and we think, legally binds the Board and Council. That the Trustees . . . should seek by resolutions to abridge our powers, is clearly against the principles of law and morality."[9] In contrast the trustees never acted as if they thought the January arrangement was a binding contract. They understood the agreement to have balanced the dismissal of Peters with the promise that Gould and his assistants would bring the observatory into operation immediately. They would be shocked by the scientists' interpretation, asking if those gentlemen could be sincere or honest in pretending that the arrangement was legally binding.[10]

In support of the trustees, the January resolutions explicitly stated that the trustees were the legal guardians of the Dudley Observatory and, as such, claimed "undivided and entire control over its property, the appointment of its officers, and its general policy."[11] Except for their insistence that references to the Coast Survey be deleted, the scientists had not questioned the wording of the trustees' resolutions. As far as the trustees were concerned, the central condition for dismissing Peters and retaining Gould was that the observatory should "be immediately placed in operation." It is obvious that a major misunderstanding existed. Apparently, the scientists who gathered in Frazer's home in February to study the resolutions assumed that the trustees were content to retain a nominal role, occupying themselves only with raising money, leaving everything else in the hands of the Scientific Council. They took it for granted that the trustees would be quiescent, content to follow the scientists' leadership, once the evil influence of Dr. Peters had been eliminated.[12]

As Bache reasoned, the endowment was, in effect, mortgaged for at least two years to recoup money already spent. The trustees would be paying no salary to Gould or his assistants. Therefore, in Bache's opinion, the trustees could exercise no decision-making power over anything affecting the operation of the institution or its director. Only decisions appropriate to their roles as trustees—raising money and approving expenditures—were expected. That the trustees would continue to have opinions and an active interest in the observatory seems not to have loomed as a potential source of problems on the scientists' horizon. As Gould later stated his impressions, the January meeting had "put an end to all sources of discord." Yet the astronomer went on to reveal that the personal animosities of January had not disappeared as far as he was concerned.

> It is true I now knew something of the two men [Armsby and Olcott] with whom I was chiefly to occupy official relations; but this very circumstance made me the more confident of my ability to avoid any new occasion of discord. By a policy of unvarying courtesy to Mr. Olcott and Dr. Armsby, I fully expected to secure the return of more congenial feeling. It was not for me to sit in judgment upon their moral character; . . . but it was my duty to rescue the Observatory from its threatened downfall.[13]

Gould was urged by his friends to pursue a quiet and unobtrusive course in Albany. He arrived to find Dr. Peters still residing in the observatory. Although received politely by Olcott, Gould returned to Cambridge to wait until Peters removed himself. Peters finally departed on 13 February; Gould and three assistants stepped off the train on 20 February. On his arrival Gould was annoyed by a series of comments in both the Albany and New York papers about his slow pace in bringing the observatory into operation. He attributed these to Dr.

Armsby but resolved to follow the advice he had received in Philadelphia. He would "pursue the path of duty quietly."[14]

From this point forward in the story, many of the pieces in the Albany puzzle depend on private reactions. What happened when people who disliked each other so much met face to face? That peculiarly Victorian sense of how gentlemen ought to behave becomes important in understanding why feelings were so bitter, but superior glances or a sneering tone of voice are often intangible. Only the resultant anger and distrust are evident. It is clear that the distrust grew on each side, each day. Events that followed Gould's arrival in Albany indicate a progressive decline in personal relationships among the principal characters during the spring of 1858.

Many of the trustees—Olcott, Armsby, March, Wilder, Pruyn—had good reason to dislike or distrust Gould before his arrival in Albany. But James Hall believed that, no matter what their feelings, they were committed to the success of the observatory. As Paul Goodman has observed, the commercial elite of mid-nineteenth century America had high standards of personal responsibility. The ideal businessman was as committed to his charitable and philanthropic enterprises as to his pursuit of career or wealth. This ethic of stewardship demanded a lifetime of integrity in all commitments, through both good and bad times.[15] If Gould had reassured the trustees that he was committed to the observatory and would bring it into operation, then perhaps he could have surmounted the obstacles between them. Gould's character, as perceived by the trustees, was a crucial determinant in the events that led to the final breakdown. Unfortunately, Gould's attitude hardly inspired confidence. He had made it abundantly clear during the previous two years that he had no desire to come to Albany. He was only present because the trustees had, in a sense, forced him to come by threatening to turn over the observatory and the promised heliometer to the hated Dr. Peters. The situation was already difficult when the astronomer arrived from Cambridge.

To put it simply, Gould was not a peacemaker. Contemporary descriptions of his personality refer to an "arrogance of demeanor toward his equals, and an overbearing conduct towards his assistants," qualities not conducive to pouring oil on troubled waters. Friends described his temperament as fiery; nor was he accustomed to weighing his words. In addition to these personality traits, Gould also had an intriguing facility for mimicry and "mordant characterization."[16] Comical mockery of men who took themselves so seriously would be deadly when directed at the ever-so-solemn cultural elite of Albany. The sharp wit of Cambridge parties and Lazzaroni gatherings could easily puncture the staid gravity of self-made and often self-important Albany merchants, physicians, and bankers. Such sallies produced short laughs at the expense of individual trustees whose foibles were well known within the community, but deep hatred would be the price when they heard of it. The Lazzaroni's worst epithet—"old fogey"—had been used so often by Gould that it would be difficult for him to brook opposition from less intellectual but equally arrogant men with opinions differing from his own. The situation in Albany demanded qualities of tact and diplomacy from the first director of the Dudley Observatory. Unfortunately, those were qualities Gould lacked.

The trustees later insisted that Gould was openly hostile from the moment he arrived in the city. They accused him of attacking and insulting members of the board and spreading the rumor that several unnamed but not unknown trustees consistently hindered his work. Gould, on the other hand, felt that Armsby and Olcott pursued a systematic course of annoyance from the time he stepped off the train, covering it with a concealing facade of respect and confidence. These later descriptions of February feelings are, of course, colored by the hot tempers of the conflict in which they were written.[17] It is difficult to ascribe blame for "opposition" or "hostility." Personal animosity toward Gould was certainly present among some of the trustees. Gould, on his part, seemed at times close to paranoiac about the constant plotting he attributed to Armsby and Olcott. Did the trustees follow a concerted campaign of obstruction that would hinder the observatory? Such a policy would work against their principal aim to have the observatory move forward as rapidly as possible. Did they trust Dr. Gould? That was a different question entirely. After the bitter January meeting, neither Gould's attitude nor his behavior induced confidence.

No matter what Gould's feelings about his situation, once he unpacked his bags at the observatory, he settled down to serious work. He planned a catalogue of all stars visible to the naked eye between the parallels of 50°N and 5°S, with an exact determination of their magnitudes to the nearest tenth of a unit. To accomplish this, he worked out the methods for a series of photometric investigations that he later continued at the National Observatory of Argentina in the 1870s. Gould immediately set his four Coast Survey assistants—George M. Searle, James H. Toomer, Erving Winslow, McLane Tilton—to work on the naked-eye determinations. Each portion of the sky was examined by at least two observers and any discrepancies were resolved by a third observer's findings. Over six thousand observations for magnitude were made, with an average of three for each star. Gould also attempted to determine the differences between observers in estimating magnitudes, resolving the average discrepancy of each from a scale he devised. Gould developed a standard of stellar brilliancy using tenths of magnitudes in Albany, which he later verified in Argentina.[18]

Bache described this work as being of great importance to science. He complained to Peirce that the trustees had little idea that while they were harassing Gould he was "making the fame of the Dudley and theirs in the eyes of the world."[19] Unfortunately, many of Gould's notes and computations were lost or destroyed. Only his conclusions were saved. These were later printed under the name of "Working List" for the stellar region surveyed. It is uncertain whether the entire body of work was completed at the Dudley Observatory or elsewhere, after 1858. Gould insisted that the work was substantially completed in Albany, and other members of the Scientific Council supported him. Dr. Peters, not an unbiased observer by any means, was equally insistent that the number of observations Gould claimed to have made in the spring of 1858 were highly unlikely, considering the number of clear viewing nights.[20]

While Gould and his assistants made their nightly observations, construction continued during the day. But if Gould expected to reestablish relations quickly with the alienated trustees, March must have opened ominously. The board met on 2 March 1858; Thomas Olcott was appointed president, replacing Stephen Van Rensselaer IV. Ample evidence of the strained relations between the director and the trustees is obvious in that the astronomer did not even know the meeting was taking place, although the trustees had met to consider Gould's requests for funds. During this meeting the trustees decided to lay down some rules about the fiscal management of the observatory. These new rules were tangible evidence that Dr. Gould's disclaimer of financial responsibility for any expenditures had changed the board's relations with him.

Prior to the conflict over Peters, Gould had been in the habit of writing drafts on individual trustees for purchases he made for the observatory. He followed an informal arrangement in which he advanced his own money and was then reimbursed. His January disclaimer denying responsibility for expenditures made up to that time made the trustees reticent to continue this more informal style of management. If Dr. Gould was not willing to acknowledge his accountability, procedures would have to change. The commercial elite who composed the board had many ties to other businesses and voluntary associations which rested on the assumption that they acted as responsible agents in the affairs they managed. If the trustees were to be blamed for extravagance and poor management, they now demanded that they be consulted before financial commitments were made. In this meeting the trustees decided that no expenditures could be made or any liabilities incurred without specific authorization from an executive committee of the board.

This caution was not inappropriate. Later in the conflict, the Scientific Council did indeed attack the trustees for sloppy management of the observatory's finances, especially with regard to informal authorizations for purchases. The Scientific Council raised many questions at that time about the "rules adopted for auditing and paying accounts," despite the obvious fact that

Gould had insisted on these informal practices.[21] Notwithstanding the trustees' stipulations about future expenditures at this March meeting, Gould continued to advance his own funds through the spring for items he thought necessary and felt injured when repayment was slow, outraged when it was refused. The appointment of an executive committee was an attempt by the trustees to set some rules on expenditures. The trustees wanted to give the endowment time to return to its full strength. Gould had been warned in January that only a few hundred dollars were available beyond the legislative appropriation for the longitude determination. His continuing habit of advancing his own funds without prior approval to purchase what he thought necessary can be judged in two ways. If Gould were trusted, the trustees might believe that he had the interests of the observatory at heart and was trying to speed things along or snap up a bargain for the good of the institution. If he were not trusted, the same act might be perceived as another example of his arrogant refusal to recognize their authority. In this already tense situation, his actions took him down paths guaranteed to generate conflict.

Once the new rules were established, the trustees turned to Gould's request for additional funds to maintain the Coast Survey field party at the observatory. The trustees were being asked to pay all the living expenses of the field party for the duration of their stay at the observatory. Again, this question can be looked at from two opposing perspectives, each leading to opposite conclusions about the motives and intents of either side. As far as the trustees were concerned, they had stressed in January that the limited funds available should be put toward setting up the telescopes. Peters had, after all, offered to remain at the observatory at minimal cost to mount the meridian circle. He had been dismissed to placate the Scientific Council. Why should the trustees now pay additional sums for food, fuel, furnishings, and stationery for the Coast Survey field party? If that same survey group were in Georgia or Alabama, they would not expect to be housed at someone else's expense. The trustees assumed, correctly, that Gould and his assistants were receiving full salaries and field expenses from the Coast Survey. In return for mounting the telescopes, the Coast Survey had the free use of all the facilities of the Dudley Observatory as well as the added incentive of the superior observations that could be made once the telescopes were set up. Why should the trustees pay all the normal living expenses for Coast Survey employees? That was not part of the initial bargain nor of the January agreement.

From Bache's and Gould's perspective, the Coast Survey was doing the trustees a large favor by establishing a field party at the Albany observatory in the first place. Gould felt that he and his assistants were performing a selfless act out of their devotion to science by voluntarily working to bring the observatory into operating condition. Simple gratitude demanded that the work force be housed and fed so as not to place any burden on the Coast Survey. If the Coast Survey had expenses for the Albany installation, questions might be asked in Congress. The very idea that the trustees who benefited from this arrangement would begrudge Gould's assistants a living allowance during the process of installing the instruments seemed preposterous.

The trustees were not convinced by Gould's argument. In response to Gould's requests for a full living allowance for himself and his staff, the trustees authorized an appropriation of $450, $300 of which was to be applied to the expense of setting up the meridian circle and $150 toward the installation of the transit. Gould was requested to mount both instruments as soon as possible. The test of Gould's intentions was clear: set up the telescopes first, and then we will talk. The trustees consistently maintained that until some practical results were obtained from the Dudley Observatory, additional contributions to the endowment were difficult to find. In their January agreement with the Scientific Council, the board had specified that Gould would be named director on condition that the instruments were set up immediately. This $450 appropriation provided the astronomer with the means necessary to meet that primary condition. Speedy execution of that task would be evidence of his good faith to the now cautious trustees.

The problem stemmed from the distinct perspectives held by

the actors in this drama. Were the members of the Coast Survey acting as volunteers? Gould certainly saw himself as a noble volunteer, working to bring the observatory along at great personal sacrifice which included leaving his home in Cambridge. As director, he assumed the trustees would provide his living accomodations. But on their side, the trustees felt Bache had agreed in August of 1855 and January of 1858 that he would maintain a group of paid observers at the observatory, justifying it by the Hudson River triangulation work and the opportunity of using the observatory's instruments. After the dismissal of Dr. Peters, the trustees understood that the Coast Survey had agreed once again to contribute a staff, supported by their regular salaries, in return for use of the existing facilities and observations obtained once the telescopes were mounted. The remaining money from the state appropriation could then be put toward setting up more of the instruments. The trustees did not consider Gould to be a selfless volunteer. On the contrary, they considered him to be a salaried member of the Coast Survey, now established at the Dudley Observatory. He had, after all, resisted coming to Albany as director for several years and still resisted giving up his Coast Survey connection to serve the observatory full-time. In effect, the trustees reasoned that they were being asked to pay for the accomodations of the Coast Survey crew that Bache had agreed to supply freely as his part of the bargain.[22] When Gould received the trustees' reply, he told them that their plan to set up the telescopes immediately was ill-advised and "ill-adapted to promote the true dignity of the Observatory."[23]

Patience was clearly needed, but the situation was not one that encouraged the virtue. From all indications the trustees were being kept uninformed of the careful, time-consuming labor involved in setting up the meridian circle. Gould's plans for setting up this telescope were intricate. Piers supporting the excellent telescopes at the Harvard Observatory had been set in place by William Cranch Bond with much less trouble and, although smaller, provided unimpeachable stability. Gould's piers were massive, seven-and-a-half tons each, and the process he devised for setting them in place was extraordinarily complicated. His dissatisfaction with the usual methods led him to construct a large derrick of his own design. The final cost of this crane was eight hundred dollars, four times what the masons had agreed to charge to put the stones in position. After the stones were in place, they had to be cut to exact height, carefully drilled, chiseled for the telescope's bearings, then clothed with felt and cased in wood. He began unpacking the eleven crates containing the meridian circle in early March. Preparatory work continued until late fall. During this time little information was passed to the trustees concerning the work on the meridian circle. All was shrouded in secrecy. Whatever his procedures, Gould erred in refusing to inform the trustees of his plan. His habitual secrecy and his unsatisfactory replies to their inquiries simply increased their suspicion and hostility.[24]

It is in a misunderstanding such as this over the meridian circle, an uninformed demand to set up the telescope at once and an unexplained refusal, that the impact of personality can be seen on both sides of the Dudley Observatory controversy. Gould later described this insistence that he set up the meridian circle immediately as a ludicrous example of the trustees' intermeddling and "excessive ignorance."[25] On their side, the trustees felt they had been patient since August of 1856. They had dismissed Peters in order to hasten work at the observatory. The explicit condition on which Gould had been permitted to remain in charge of the observatory was "the *immediate* mounting of these neglected instruments." Yet as far as they could see, despite their repeated requests for information on the telescopes, Gould continued his delaying tactics. In a meeting three days later, on 9 March, the trustees voted that all requests for funds must be submitted in writing, and that all money requested for setting up the instruments would be provided, but that was all.[26]

The meridian circle remained a focus for tensions and increasing irritations throughout the spring of 1858. Gould was not only secretive about its progress in installation but announced that he had sent off sections to an American optical craftsman for refitting, thus invalidating any claim for possible imperfections against the makers, Pistor and Martins of Berlin. Gould also sent the German craftsmen an unsolicited and unauthorized gratuity of three hundred dollars when they complained they had underestimated the telescope's cost and suffered a loss. Gould insisted the gratuity was warranted and asked the trustees to repay him.[27]

Spring was in the air by the end of March, but relations between Gould and the trustees remained icily formal. Petty disagreements eroded patience and good will on both sides. The divergence in perspectives is again obvious in these misunderstandings. One example of the irritations arising from this difference in perspective can be seen in the bad feelings that arose over the question of hired help at the observatory. Gould was convinced that Joseph McGeough, the Irish caretaker who had resided in the basement of the observatory with his family since 1852, was Armsby's spy. While he waited for McGeough to move, Gould hired his own housekeepers, the Bygates, at a salary of seven hundred dollars per year without consulting the trustees. To Gould, dismissing the slovenly and "drunken Irishman" and hiring Mr. and Mrs. Bygate as housekeeper and caretaker were necessary measures for bringing the observatory into professional working condition. He described the reluctance of the trustees to pay the Bygates's salary as a "specimen of the kind of annoyance with which Armsby and Olcott daily amuse themselves. No petty attempt to irritate me is too low or contemptible; no unprincipled manoevre too mean for them."[28] As far as the trustees were concerned, Gould had no right to commit the observatory to a financial obligation of seven hundred dollars per year without prior approval. They had been explicit about the limited funds available at this time. In their eyes Gould was once again disregarding the terms of the January agreement and the stipulations about expenditures made by the board earlier in March.

In addition to the disagreement over the Bygates, articles criticizing Gould for being slow in setting up the transit reappeared in the local papers. Gould attributed them to Armsby and Olcott as part of a campaign to force him to leave. It is certainly possible that the trustees hoped Gould would leave of his own accord. In January they reluctantly dismissed Peters. They had accepted Gould, provided he set up the telescopes quickly. By late March his repeated insistence on delay, the slow pace of work, his "excessive particularity," and his refusal to answer questions about the progress of work at the observatory sapped whatever store of patience and trust might have remained. If Gould left Albany of his own accord, relations could be preserved with the Scientific Council, and a new director might be sought who would bring the observatory into operation.

Gould had promised the members of the Scientific Council that he would follow a course of "undeviating courtesy and constant conciliation" despite all difficulties, setting 9 June 1858 as his own personal deadline. He was determined to provide no grounds for censure until then. But Gould's version of undeviating courtesy toward those he considered his enemies was not always helpful in the increasingly suspicious atmosphere. Gould had been known for his boisterous pranks as an undergraduate. As Gould described one small incident to Peirce, he solemnly saluted Dr. Armsby whenever he saw him on the streets of Albany. Lifting his hat in an exaggerated gesture of greeting, Gould undoubtedly enjoyed the obvious discomfiture of the doctor, who could not complain of discourtesy but knew he had been insulted nevertheless. Stiffly formal gestures were as mocking as the complete absence of gentility. Humorous insolence, as in the sneering doffing of a hat, simply confirmed prior impressions.[29]

Although he exhibited a facade of crisp bravado to his enemies, Gould was sadly experiencing the anxieties of another downward emotional spiral during the spring of 1858. He confided to Peirce that he was utterly exhausted. April arrived, and Gould confessed that he was finding his course of forced politeness increasingly constricting. He vented some of his irritation in a letter to the prestigious *Astronomische Nachrichten*. In describing his work in Albany, Gould referred pointedly to the trustees' refusal to provide funds for necessary expenses, announcing to the international community of astronomers that he received no salary from the Dudley Observatory and was forced to do longitude determinations in order to gain his livelihood. This was all true. What he did not say in the letter was that these were also the agreed terms on which he had accepted the position in January. The trustees had been clear about the fiscal condition of the observatory. The original agreement between Bache and Armsby had established a mutually reciprocal arrangement in which the Coast Survey would set up the instruments and, in return, have the use of them in longitude determinations. Gould presented the facts to the most influential astronomical journal in the world in a manner that made all these arrangements appear to be unexpected hardships he was forced to endure. His letter appeared on 3 May 1858. It provided yet another reason for the trustees to distrust Dr. Gould.[30]

The coming of spring meant that field work could be resumed, and the Coast Survey began preparations for the Albany longitude determination. The members of the Scientific Council had their own reasons for wanting the longitude of the Dudley Observatory established as quickly as possible. By utilizing their control of the Albany observatory and the Coast Survey, they hoped to supplant the Harvard Observatory as the prime station from which all other longitude in the nation would be derived. If the longitude of the Dudley Observatory were determined quickly, the observatory's position in the Hudson River triangulation would give it preeminence in the region, superseding Harvard's, because the Coast Survey would continually use it as a reference point. This was part of an attempt to establish the Dudley Observatory as an institution of national importance, under the control of the members of the Scientific Council, as well as a reflection of their dislike for the Bonds of the Harvard Observatory.

As far as Bache was concerned, the two thousand dollar appropriation for the longitude determination, obtained from the state legislature through the influence of General Robert Pruyn, was evidence of the state's willingness to support scientific work in difficult times. It justified his own continued involvement with the observatory.[31] Unfortunately, the appropriation became another source of contention between Gould and the trustees. The actual cost of establishing the observatory's longitude was estimated at about six hundred dollars. Gould assumed the balance remaining would be placed in his hands to be used at his discretion. But the trustees interpreted his effort to secure unrestricted use of the longitude funds as an attempt to circumvent their supervision of expenditures. They became increasingly suspicious of his "great anxiety to grasp the Legislative appropriation."[32] No one suspected Gould of veniality. It was a question of authorized procedures, generated by the questions of authority and accountability he had raised with his statements in January.

Misunderstandings developed as communication eroded between Gould and the trustees. The trustees assumed that, as with most agreements for services, the six hundred dollars for the longitude determination would be paid to the Coast Survey when the work was completed. This was normal practice in contracting for services in the 1850s, but it led to trouble with the superintendent of the Coast Survey. During 1857 the trustees had advanced funds for the deposit on the Coast Survey transit, paid the shipping costs, and purchased the piers required for its mounting. When pressed by Bache to pay cash for the observatory's longitude determination, Olcott asked if it would be possible to deduct the amount already advanced by the trustees in behalf of the Coast Survey transit from the bill for the survey's longitude work. Bache refused, insisting on full cash payment in advance, arguing that he had no authority to use funds from his congressional appropriation to do longitude work for the private benefit of the Dudley Observatory. As superintendent of the Coast Survey, Bache had defended himself earlier against criticism of the Albany deployment of survey personnel by stressing that the longitude work would be paid for and was not a donation of the survey's services to the observatory. Bache was wary of offering grounds for congressional criticism. Coast Survey field work, he insisted, could not be exchanged for freight bills and stone piers for a telescope that required a permanent installation. The survey was, after all, forbidden by Congress to have a permanently mounted instrument, even though the Coast Survey would own the transit.[33]

Obviously, Bache was treading a fine line. The Coast Survey's congressional appropriation had to be safeguarded at all costs. The Dudley Observatory must have its longitude established. Bache was intimately associated with each institution, but he insisted that his actions in one had no relation to his activities in the other. The Coast Survey field party was at the observatory installing the instruments, one of which belonged to the survey, but they were not there to benefit the observatory. The Dudley Observatory would be the prime point in the Hudson River triangulation for the Coast Survey, but the observatory must pay cash for the longitude work to establish the position that would make its primacy possible. Cash advanced by the trustees for the survey transit had no relation to the cash demanded for the determination. As ever, Bache redefined the meaning of the unofficial connection he insisted existed between the observatory and the Coast Survey to suit the circumstances.

As preparations for the longitude work in Albany neared completion, Bache refused to continue unless the six hundred dollars was paid in advance by the trustees. Gould was the man in the middle. He forwarded Bache's letter to the trustees, noting that he could not afford to advance this sum out of his own pocket. At this point one of the frustrating little minuets that made the Dudley Observatory conflict so maddening to all involved took place. Each individual in this conflict was concerned with the niceties of proper behavior. Emphasis was placed on form and, as a result, substance often suffered. The longitude survey is one such instance.

Because of the continuing atmosphere of distrust, communication between Gould and the trustees was almost completely reduced to formal letters. Bache and Gould wrote on 22 April, requesting advance payment for the longitude work. Two days later Olcott replied that he thought everyone understood that no money was on hand; the trustees were paying observatory liabilities out of their own pockets at this time. Funds from the legislature for the longitude work were not advanced in a lump sum; the legislative appropriation stipulated that the money would be provided when the work was completed. Bache wrote again four days later, emphasizing that he had no authority to do a longitude survey for the private benefit of the Dudley Observatory. The work must be paid for in advance. On 28 April Olcott wrote Gould that he had no objections to the sum demanded by the Coast Survey but reiterated his belief that the work should be paid for when completed. Gould replied the same day. He had used two hundred dollars of his own money to keep the work going but, in the absence of more funds from the trustees, the longitude field parties had all gone home on 27 April! Olcott replied to this devastating news with a short note saying that the entire six hundred dollars would be sent immediately from his own pocket "to avoid further misunderstanding." The correspondence resembles a classic gavotte—the dancers parading to and fro while they concentrate on correct postures and steps, bowing and moving around the floor until they return to their original positions.[34]

With Olcott's cash in hand, the survey crews reassembled. A telegraphic connection was set up between Lewis Rutherfurd's private observatory in New York City and Albany. The selection of Rutherfurd's observatory rather than the Harvard Observatory would be another source of recrimination between the trustees and the members of the Scientific Council as well as between George Bond and Gould. The longitude of Rutherfurd's observatory had itself been derived from that of Harvard. The use of a derived rather than original site in the determinations was attacked as both costly and unnecessary, stemming from Gould's dislike of the Bonds of Harvard Observatory. The work, once started, was hampered by telegraphic problems and weather. Gould did not publish the final report on this attempt to establish the observatory's longitude until 1861. In the meantime the longitude was derived by direct connection with the Harvard Observatory in 1860.[35]

In retrospect it is clear that relations between the director and the trustees of the Dudley Observatory deteriorated progressively between February and May of 1858. Who was to blame? Certainly faults have to be apportioned about equally on both sides. Money was tight. From the trustees' perspective, the treatment of Dr. Peters by the scientists seemed both arrogant and shabby. Dr. Gould's aversion to taking up a position openly associated with the Albany observatory lessened local trust in his commitment to the institution. Gould's January disclaimer of any responsibility for prior expenditures lessened the trustees', especially Olcott's, faith in his integrity. The central standard of the antebellum commercial ethic was personal responsibility.[36] When Gould apparently reneged on his responsibilities, faith in his character diminished accordingly. Dr. Armsby was an officious busybody whose energies, when directed in appropriate channels, could be tremendously effective. That same tenacious energy could be tremendously irritating in a situation where the physician not only distrusted someone but felt personally betrayed. Critical eyes, insisting he prove himself, were turned on Gould when he was finally forced to come to Albany. In this suspicious atmosphere, Gould became more defensive and even more secretive. He resented all interference, especially from Dr. Armsby, and would not answer questions he considered to be ignorant about work going on at the observatory. Gould's attitudes fed doubts about his trustworthiness and generated increasing uncertainty about his character and intentions. As a result, when hostilities finally came to the surface, there were no reserves of faith or trust in good intentions upon which either side could draw to avoid a final rupture.

CHAPTER 8

The Final Straw

By early May of 1858, Gould's friends were commenting in private on his involvement in greater numbers of arguments as well as on his low emotional state. Bache urged him to concentrate on science instead of dwelling on minor irritations, but it was increasingly difficult for Gould to do so. He felt he was being intentionally harrassed with annoyances and personal indignities but insisted he would preserve his self-respect. As he put it, he had resolved that the observatory "should not fall a victim to the bad passions of unprincipled men, though this might be the case with myself; and I therefore strove with all my powers to conciliate and soothe these men—who apart from their relations to the Observatory, would of course have been unworthy of a thought."[1]

Gould especially resented continuing inquiries about the installation of the telescopes. He responded to such questions in what he described as a dignified manner, but the trustees thought the replies were "evasive—sometimes insolent, and never satisfactory." A growing number of the trustees were increasingly of the opinion that the astronomer had no intention of bringing the observatory into immediate operation or into a condition that would inspire public confidence. Even Gavit, Gould's loyal friend, was rebuffed when he urged Gould at least to set up the Coast Survey transit, reporting to Bache that the astronomer had became very unpleasant when he persisted on the subject.[2]

Tensions were clearly building on both sides. As small issues were raised in the studiedly formal letters that passed back and forth between the director and the Executive Committee during the month of May, tempers were held by sheer force of will. Any small incident, like a spark on dry pine shingles, could lead to an explosion out of proportion to the initial cause. Despite good intentions on both sides, the last tattered remnants of cooperation between the director and the trustees shredded beyond repair on 19 May. That afternoon, Dr. Armsby and John Wilder arrived unexpectedly at the observatory. They had come, they said, to inspect the railroad line that ran at the base of the hill on which the observatory stood. Once they were at the observatory, the two trustees decided to look around the grounds and buildings since so many new expenditures had been proposed by the director. What followed was a farcical series of affronts to "face," in the Oriental understanding of the term.

D.E. Winslow, one of Gould's young assistants, saw Armsby and Wilder approaching. He quickly locked all the doors to the observatory building, following Gould's orders, so that no one might enter "without proper permission and attendance." Wilder and Armsby met another of Gould's assistants, James Tilton, and on impulse (according to their account) decided to enter the observatory. The two trustees asked Tilton for the keys in a manner the young man found offensive. He replied that he did not know where they were. Wilder then repeated the question in what Tilton described as an "indecent" manner. Tilton, being a young man of spirit, replied heatedly. Both trustees then continued on to the observatory, with Tilton hustling along behind, insisting that they seek Dr. Gould's permission to enter.[3]

While tempers were fraying, Gould suddenly appeared. Although Dr. Armsby maintained that Gould had actually been present at the observatory all the time, the astronomer insisted that he had been on his way to an appointment in town, forgotten something, and returned to find Armsby and Wilder trying to enter the observatory in his absence. His feelings for Armsby were already vitriolic, and Wilder had been one of Peters's strongest supporters. Hurrying across the grounds, Gould confronted two trustees whom he described privately as among his worst enemies. In a stiffly formal manner, the astronomer stated that although he was already late for an appointment, he would escort them through the observatory if that was what they wished. The trustees declined, as gentlemen, to impose on his time in this way as he was obviously in haste to leave. They later said they did not wish to delay him for fear he would use the incident as another example of his "bed of thorns" in Albany. With their refusal, Gould left.[4]

As Gould went down the hill, Winslow took the craftsman working on the clock system into the observatory. At this point Armsby and Wilder apparently decided to follow, "on the spur of the moment" as they later described it. The building was being unlocked, and two men were obviously entering. Why not enter with them? They followed Winslow up the steps. As they reached the entrance, Winslow turned quickly, slammed the door in Armsby's face, and locked it. Faced with this affront to their dignity, the two trustees now loudly demanded entrance. One of the young assistants standing outside offered to go after Dr. Gould to obtain his permission for them to enter the observatory, but Armsby and Wilder indignantly refused this added humiliation. They now insisted on being allowed to enter on their authority as trustees, without reference to Gould. The young men refused. Gould's assistants were wary of taking responsibility for any gesture recognizing the trustees' authority over the institution. Entry to the observatory had obviously become symbolic for everyone concerned. The question of authority that had been a vague source of irritation was now clearly and explicitly a divisive issue.[5]

Dr. Armsby and John Wilder continued to knock loudly on

the doors of the building, but Toomer, Winslow, and Farmer just continued working inside, making audibly disparaging noises as they did so. Enraged at being "sneered at by the young men employed there," Armsby and Wilder had what can only be described as a temper tantrum. They peered in the windows, continued pounding on the doors for a long time, and behaved "in a manner neither dignified, gentlemanly, nor decent."[6] It was not an afternoon that would increase the trustees' affection for their director or for his young assistants. The sneers of the young men were interpreted by the irate trustees as an obvious reflection of Gould's own contemptuous attitude toward the board. Their refusal to allow the trustees to cross the observatory threshold could only be sustained if ordered by the director himself. Given the events of the previous year, it is easy to see why the question of access to the building became symbolic for everyone.

When he returned, Gould listened to his assistants' version of the afternoon's events and praised them for what he described as their responsible actions. He instructed them to write up reports of the whole business for the Scientific Council.[7] Gould remained at the observatory for two days, then left again on Friday 21 May. As luck would have it and apparently without prior planning, General Robert Pruyn and Dr. Armsby chose that time to arrive once again without warning.

Robert Pruyn was a trustee. He was also speaker of the State Assembly. He maintained that he had been sent by the trustees to the observatory to discharge a duty. Gould later charged that Pruyn had, in reality, come to discuss a bill from his iron foundry for stone drills used in preparing the piers for which Gould refused to authorize payment. Whatever Pruyn's reason, Dr. Armsby had been invited to accompany him. When they arrived at the observatory, the two trustees met Winslow and asked to be admitted. Winslow refused, saying that Gould was once again absent. No one could enter. Pruyn pressed the young man hard, claiming admission as a trustee, and Winslow went off to find the keys. Winslow later wrote that he did not wish to precipitate another incident, but he infuriated the two trustees by leaving them to cool their heels on the steps for over forty minutes. The keys he brought with him only opened the main door and the empty transit room of the observatory, leaving all the other rooms securely locked.[8]

After a cursory look around, the two trustees turned to leave. As they went down the steps, Winslow attempted to explain why he had slammed the door in Armsby's face on his previous visit, but General Pruyn forcefully interrupted. Pruyn suggested strongly that the young man make certain of the precise limits of his authority as an assistant in the future. Winslow responded just as forcefully, saying that the trustees had to seek permission from the director before they would be allowed to enter the observatory building. Winslow was insistent that "tho' this matter was but a form, yet therein lay the difference between recognizing you [Gould] as Director of the Observatory and refusing to do so." Pruyn replied "with much temper" that he had no wish to call upon Gould and that it was not Winslow's place to dictate to him.[9]

James Tilton, another of Gould's assistants, hurried to meet the trustees, finding them as they stood on the observatory steps exchanging heated words with Winslow. Tilton described Dr. Armsby as being "in a state of mind incompatible with my ideas of his disposition." Armsby demanded an account from Tilton of his actions on the previous day, especially wanting to know if Gould had been present when Tilton said the astronomer had left for an appointment. Armsby believed that Gould had been there all the time, with no pressing appointment in town at all. Tilton "answered evasively." Tilton's report to the Scientific Council implies that Armsby was correct, that Gould had been present on Armsby's prior visit. Thoroughly disgruntled, Pruyn and Armsby decided to leave. Thinking perhaps to catch Gould in the same subterfuge, Armsby stopped to peer through the windows into the back parlor of the dwelling house as they went past. Winslow opened the door opposite, catching Dr. Armsby in the act of peeking in the windows. Dr. Armsby was not only angry but now highly embarrassed at being caught in such an ungentlemanly position.[10] With such histrionic outburts and classicly comic gestures, the action at the observatory assumes the broad humor of farce. The doctor was obviously overwrought with temper and frustration. Unable to endure the goading provocations of these two visits, he was reduced to a burlesque-like caricature worthy of Gilbert and Sullivan's lighter lyrics. One can easily imagine the four assistants as the chorus in such an operetta, full of witty asides and musical puns as the melodrama unfolded.

Once again Gould listened to a tale of the trustees' humiliation. His assistants—Toomer, Tilton, and Winslow—described Armsby and Pruyn's language and conduct as "offensive and dictatorial."[11] Gould complimented the young men on their self-control and manly spirit, reiterating his rule that all visitors must seek his personal permission to enter the observatory while he was on the premises. In the event of his absence, his assistants had full authority to decide whether a visitor would impede any scientific work in progress.

Both sides were now intent upon preserving their understanding of their legitimate authority within the institution. Both sides saw the incidents as part of a concerted campaign by a hostile opposition to diminish that legitimate authority. Gould believed the visits to the observatory "were unquestionably part of a systematic effort" on the part of the trustees to humiliate him.[12] On the trustees' side, rude treatment by Gould's young assistants offended their sense of the proper deference due to gentlemen in their position, men who were legally responsible for the institution. The treatment received by Dr. Armsby, General Pruyn, and John Wilder offended them. There had been earlier reports of others who had been treated discourteously by Gould's assistants, to the detriment of the observatory's reputation. Because the young men were acting on Gould's orders, the trustees believed that Gould intended to treat everyone, trustees included, in a contemptuous manner and keep them completely in the dark about the observatory's progress. Dr. Gould, as far as the trustees could see, wished to establish his own authority as completely superior to theirs in all questions.[13]

The trustees were not refusing to accept Gould's professional authority as an astronomer. They were rejecting his refusal to acknowledge their own legitimate role within the institution. He refused to give them any information and barred them from the premises. His habitual arrogance and impatience exacerbated a difficult situation. In contrast, the Bonds at Harvard Observatory enjoyed cordial relations with their visiting committees, enhanced by a sense of mutual trust and open communication. The Visiting Committee for the Harvard Observatory arrived at appointed times and was given full reports, but members of the committee often visited as individuals, asked questions informally, and received immediate replies. By refusing to give the trustees any information about what was going on at the observatory, Gould escalated an accidental question of access into one of absolute authority. The trustees had been generous in their commitment to the observatory. Gould's rejection of responsibility in January, the lack of trust between the trustees and the astronomer, the animosity between several trustees and Dr. Gould, his secrecy—all combined to make this question of access seem like the final straw.

The Executive Committee met on 22 May 1858 to consider the two incidents. They were told that members of the board of trustees had been grossly and deliberately insulted by Dr. Gould through his youthful subordinates. Members of the board, "in the discharge of their duties," had been "treated with insolence, prevarication, and falsehood."[14] In response the trustees adopted a series of resolutions affirming their authority.

The observatory grounds must be open from seven in the morning until seven in the evening every day but Sunday. Although this is uncommon today, many nineteenth-century observatories allowed visitors to stroll around their grounds. Certainly, the college observatories did not restrict the use of the surrounding grounds. Before Gould's arrival in February of 1858, the grounds of the observatory had been open to "all well mannered and well intentioned persons." The trustees also insisted that duplicate keys to the building must be provided for the sole use of the trustees. This was an obvious response to the trustees' confrontations with Gould's assistants. Finally, Dr.

Gould was ordered to instruct his assistants to be both courteous and civil. They reminded Gould that the observatory had been endowed through the generosity of the citizens of Albany and was still dependent on the continuation of that good will for its future success. These resolutions were carried by hand to Gould that afternoon.[15]

On receiving the resolutions, Gould immediately sat down to write the members of the Scientific Council. He enclosed copies of the resolutions, the reports from his assistants, and his own endorsement of the young men's actions, saying that they had "behaved like gentlemen and men of discretion." The trustees' actions, he said, were "of course incompatible with the success of the Observatory." In a cover letter to Peirce, Gould insisted that the time had arrived for the Scientific Council to take a stand. Gould believed that Armsby and Olcott were attributing his conciliatory manner to a "want of spirit" and were intent on reducing him to a menial position. By refusing to obey the trustees' rules, Gould feared that he and the Scientific Council with him would be forced to give up the observatory. If the whole council would take a stand behind him, he declared he was ready to "stay by the colors . . . tho' it cost 10X [sic] my life."[16]

While he waited anxiously to hear from the other members of the Scientific Council, Gould retired into his observatory shell, disturbed only by his sensitivities to the rumors and gossip floating around Albany. Despite a brief and seemingly cordial visit with Thomas Olcott and John Rathbone on the evening of 25 May, Gould's isolation and his fears increased. He was convinced that the trustees, especially Armsby, were conspiring against him. Isolated as he was, both physically and emotionally, it is not surprising that Gould felt apprehensive and vulnerable. His pride insisted on a strong and manly posture, but his letters reveal his increasing fear that he might be ignominiously humbled.[17]

In the meantime Bache and Joseph Henry conferred in Washington. Bache had his own problems. The secretary of the treasury was asking specific questions about Coast Survey expenditures, and the survey's budget was under attack in Congress by members from western states. The vexations at the Albany observatory were a "nasty business" to have to deal with at this difficult time. Bache advised Gould to restrain his temper, counseled patience, and attributed all of Gould's problems to "one obstructive individual." Bache most likely had Dr. Armsby in mind. Bache and Henry drafted an official reply to Gould in which they supported his rejection of the resolutions.[18]

Public access to the grounds was absurd; access by the trustees was equally impossible. In their opinion the trustees only had the right to visit the observatory officially as a body. As individuals, trustees had no such rights. If the building was not under Gould's control, they argued, neither were the instruments. The scientists contributed their own dash of hyperbole by comparing the trustees' demand for a set of keys to the "extreme claim of Imperial police agents." The Scientific Council approved Gould's actions completely and urged the trustees to rescind their foolish resolutions. If the trustees should refuse, then Gould was to appeal officially to the Scientific Council, who would then pass judgment on the question. Later, the Scientific Council would describe the trustees' resolutions as unprecedented and reprehensible examples of meddling. None of the members of the Scientific Council addressed the essential question of why the two confrontations had arisen in the first place.[10]

In the Dudley Observatory, the boundaries of authority had never been precisely defined. The institution took its strength in 1856 and 1857 from the mutual sense of trust prevailing among those involved in its affairs. The observatory developed in an atmosphere of friendly but decidedly informal relations between a scientific elite and a cultural elite. In such an unstructured situation, ultimate questions of authority and disputes over the perimeters of power would be the kiss of death to a fledgling institution not yet able to support itself. The Scientific Council's unqualified endorsement of Gould's claim to what appeared to be an extensive authority and their approval of the behavior of his young assistants threatened the fragile foundation of trust upon which a new institution depended. More significantly, the scientists never addressed the question of the responsibility of a director of such an institution to inform a board of trustees of his progress or plans. Given the prevailing expectations with regard to the roles of trustees in antebellum institutions, the Scientific Council's full endorsement of Gould and his assistants' behavior was unquestionably impolitic in 1858, especially after the two traumatic incidents at the observatory had heightened tensions so drastically.

On the last day of May, Gould heard that a meeting of the Executive Committee was to be held. He wrote to say that if the trustees insisted that the grounds of the observatory be used as a park, a fence must be built and a patrol hired to guard the premises. He described restrictions imposed by the directors of European observatories and argued that his responsibility for the instruments required "exclusive custody and control" that even the trustees could not override. Gould then poked insensitively at the observatory's fiscal condition. This was an area in which some of the trustees, especially Thomas Olcott, were understandably touchy. Gould wrote that, despite the fact that the trustees had spent all the available funds, the Scientific Council had generously agreed to maintain their connection with the institution. The trustees, of course, had their own opinions about who had been responsible for spending the observatory's money. Gould then referred to the gratuitous labor given by him and his assistants to the observatory, for the sake of science. He argued that such generosity entitled them to be free from interruptions that might jeopardize their work.

Again, the trustees' perspective differed. The so-called interruptions were attempts to seek information about what had been accomplished at the observatory, information the director refused to give them. In their minds Gould and his assistants had promised, as part of the January agreement, to set up the telescopes. If they were not setting up the instruments, why were they at the observatory? They had the use of the site and its facilities for their Coast Survey work in return for setting up the telescopes, which they were free to use once installed. Gould refused to provide any information about what was going on. The trustees were legally responsible for the institution. From their viewpoint they were not being unreasonable in attempting to determine what progress was being made toward bringing the observatory into operation.

Gould strained tempers further by endorsing the behavior of his assistants. The astronomer strongly reaffirmed his approval of their actions, describing their conduct as manly and correct in both incidents at the observatory. The young men, Gould wrote, deserved gratitude and approbation, not an unjust censure from the trustees. Gould refused to admonish his assistants and ended with an emphatic refusal to meet with the trustees to draft rules for admission to the observatory.[20] Gould's letter contains many references to proper gentlemanly behavior. He was unquestionably aware of the deep affront taken by the trustees from the attitudes and conduct of his assistants. The trustees included some of the most respectable and influential citizens of Albany—a city known, like Boston, for its decorum as well as its cultural patronage. From the perspective of the trustees, Gould was not only refusing to recognize their rightful authority in the institution but was also condoning disrespectful conduct toward members of the board of trustees.

The anger that Gould's letter induced is obvious. The message was described by board members as "arrogant, insolent, and unbecoming the position held by Dr. Gould and his relation to the Trustees." Thomas Olcott later described Gould's letter as "sufficient justification for terminating relations in itself." The trustees were perplexed at Gould's assumption that he could tell them what they could and could not do in an institution for which they were legally responsible. They described him as putting on airs, reviewing and condemning the trustees' actions, and refusing to pay any attention to their resolutions.[21]

Gould later confessed himself surprised at such an intemperate reaction to a letter he described as "respectful, courteous, and as conciliatory and kindly as self-respect or manliness permitted." The Scientific Council defended Gould's letter as a "model of plain, direct, manly, independent statement," and charged that only tyrants objected to such plain speaking.[22]

Again, perceptions made all the difference. What was manly to Gould was perceived as arrogant and insolent by the trustees. Thus, his refusal to correct his assistants was a final, intolerable gesture of contempt. In an increasingly tense situation, refusal to recognize such heightened sensitivities was certainly impolitic on the part of both Gould and the members of the Scientific Council.

A full meeting of the board of trustees convened on 4 June to consider Gould's letter. The discussion was heated. Stephen Van Rensselaer IV, who had been involved in decades of litigation and well-publicized disputes after his father's death, submitted his resignation from the board, stating that he was unwilling to risk being drawn into any more public quarrels. Ira Harris, justice of the New York State Supreme Court, was named to take Van Rensselaer's place and took his seat immediately. Gould described Harris as the "strongest Peters-ite in Albany" and disliked him intensely. The trustees continued in a stormy and turbulent session.[23]

When Gould's letter was read aloud to the assembled trustees, John Wilder angrily condemned it. Wilder had strongly supported Peters and had also been involved in one of the incidents at the observatory. He especially resented Gould's approval of his assistants' conduct. Wilder first questioned Gould's right as director to refuse orders from the board of trustees and then reduced the question of keys to a fundamental statement about authority within the institution. Access to the observatory grounds and buildings by members of the Board of Trustees, Wilder insisted, "has become a question as to where the supremacy of this institution is vested and must now be determined."[24]

Relations between the trustees and the director had come to the point of crisis. From the time of the Peters controversy in January, tensions had been building until they reached this final intolerable level. The trustees felt that Gould had been vindictive toward any who had supported Peters. In their minds Gould had finally surpassed the boundaries of acceptable behavior in openly condoning the humiliation of several trustees by his subordinates. A majority of the trustees felt that Gould had adopted a policy of rule-or-ruin in his refusal to answer any inquiries about the progress of the institution. His "most formal and ceremonious" behavior since his arrival was interpreted as plain evidence of an arrogantly contemptuous attitude toward trustees who had worked hard to establish the institution placed in his hands. Now, in addition to refusing any information, those trustees could not even venture onto the grounds of the observatory for which they were legally responsible unless they humbled themselves before him. Instead, they were to be turned aside, however rudely, by his assistants.[25]

Overlooking their divisions with regard to future policy, it should be noted that the *entire* board united in condemning Gould's conduct toward the trustees who had been insulted. The resolutions providing keys for the use of the trustees were also unanimously approved. The board, in its entirety, also expressed dissatisfaction with Gould's refusal to answer any questions about his progress in bringing the observatory into operation. The board of trustees did not split in their votes until they tried to decide what further course of action should be taken. They then divided into a majority of nine and a minority of four or five (depending on how many were present) on the question of severing relations with Gould, and concomitantly, with the other members of the Scientific Council.

Judge Harris offered the resolution that placed the question before the board in unequivocable terms. Because of constant difficulties "arising from want of harmony between Dr. Gould and the members of this Board . . . some new arrangement is absolutely necessary."[26] In the niceties of nineteenth-century language, Gould was fired. Olcott and the majority of the board maintained that Harris's resolution was adopted unanimously. John Gavit reported that the trustees divided nine to four in the vote. Whatever the final tally, the results were clear when the meeting ended at midnight. Some new arrangement was necessary. The news was carried to Gould through a rainstorm by "chivalric Gavit."[27]

The "want of harmony" resolution, as this was later known, was sent at once to the members of the Scientific Council. In a cover letter to Bache, Olcott expressed his personal gratification with the trustees' action.

> . . . yet with equal candor, I can say for myself and my associates, that the necessity which impelled our action, was the occasion of painful regrets.
>
> Looking with confidence and pleasure to the continued friendly co-operation of the other members of our Scientific Council, I have the honor to be,
>
> With undiminished regard,
> Your obedient
> Thomas W. Olcott[28]

CHAPTER 9

Who Is in Charge Here?

With the "want of harmony" resolution passed by the board of trustees, events at the Albany observatory escalated from annoying irritations into a full-scale confrontation. Unfortunately, the fireworks blazed up at a time when tempers among the members of the Scientific Council were already frayed. Joseph Henry was awash in the details of moving the various expedition collections from the basement of the Patent Office over to the Smithsonian. Bache was busily defending the Coast Survey budget from another congressional attack. Western members of the House of Representatives argued that the budget of the survey, which concentrated on the eastern states, should be trimmed severely in a time of economic hardship. To add to his woes, the Coast Survey's accounts were being audited by an unfriendly comptroller. Benjamin Peirce spent a contentious spring, engaged in angry sessions in the American Academy. Peirce resigned in a fit of anger, then relented, only to have his reapplication refused. His embarrassingly public conflicts with

C.F. Winslow and John Warner over charges of plagiarism were also generating comment.[1]

Peirce feared that Gould's course of action had created a situation in which the Scientific Council stood a good chance of being driven from Albany "in contempt." In Peirce's opinion there was little chance that the Scientific Council would recover its position in Albany; he suggested that they shift attention to the Harvard Observatory instead. He wrote John LeConte that William Cranch Bond would soon "go to the stars," as he was sixty-nine and clearly in poor health. When the vacancy should occur, Peirce planned to displace George Bond and be named director of the Harvard Observatory himself: "I shall call the Coast Survey and the Smithsonian to my aid . . . it will be a hard fight—but I hope to gain it."[2] Despite Peirce's hesitation about the Albany institution, Bache remained convinced that the future possibilities of the Dudley Observatory were worth a stand.

Bache was also highly resentful of Gould's dismissal. He described Armsby's course as a dirty one, and the trustees were "pigs of the first order." After closeting himself with Joseph Henry, Bache decided to respond to the trustees with his own interpretation of the "want of harmony" resolution. The Scientific Council would regard this as an appeal for advice rather than as a statement of intent. The council must present themselves to the public as the persons "responsible for the scientific character of the observatory." If the public could be convinced that the scientists had greater authority, the trustees would look ridiculous in criticizing the director. As Bache planned it, it would be "a scathing exposure when it comes and kill them dead, even if they reject us."[3] Bache and Henry made plans to meet Gould at the home of Wolcott Gibbs in New York. Gibbs, one of Gould's closest friends, was vehement in Gould's defense, insisting that the Dudley Observatory should be sent into oblivion rather than be taken from Gould.

The ever-present Peters still perched like a bird of ill omen, never far from the thoughts of the members of the Scientific Council as they deliberated. If Gould had been fired by the trustees, Bache had little doubt that the observatory would soon be turned over to Dr. Peters. Gould reported that Peters appeared at the observatory to have a look around the day after the resolution dismissing him had been adopted, confirming Gould's impressions of an organized conspiracy. Throughout the controversy Gould attributed many of his difficulties to Peters's dark machinations and carried his hatred for the immigrant astronomer with him for the rest of his life. The animosity was reciprocated.[4]

As the Albany situation continued to deteriorate, Peirce considered resigning from the Scientific Council. Gould protested, fearing that if Peirce resigned, "some enemy," like James Hall, would be appointed to take the Harvard mathematician's place.[5] James Hall's part in the conflict was a curious one. He supported both sides alternatively, helping and hindering as he went. Hall strongly urged the appointment of Dr. Peters in 1857 but later described the entire effort as absurd. As the controversy unfolded, the state paleontologist first railed against Gould and then later supported him. Gould was distrustful of Hall's quixotic behavior in late May of 1858, but wrote that he was certainly "playing sweet." Wolcott Gibbs termed Hall a complete rascal. Hall's biographer later wrote that the geologist could not keep his fingers out of the boiling pot, although the observatory dispute was none of his business, and he was sure to be scalded when it spilled into the public press. As bitter letters entered the newspapers in late June, James Hall worried that the hostilities might split the city to the detriment of all scientific work done there. After attempting to act as a mediator, he wrote Bache that "the spirit of parties on both sides" prevented any reconciliation.[6]

By the middle of June, any hope of mediation was long gone. The rumors and gossip that floated up to the observatory from the city below bothered Gould immensely and prevented him from concentrating. Gibbs reported that the astronomer was in a very excitable state of mind. Bache, somewhat insensitively, urged him not to be disturbed by petty irritations and, at the same time, suggested he be more prompt in his Coast Survey work. Gould complained of this lack of empathy, wondering if "the Chief" would be able to stand it himself if they changed places for a day. As it was, he intended to bear "all for the sake of science and my self respect."[7] Friends advised him to remain quiet and keep his temper, but Gould was extremely irritated by new rumors that Armsby was saying the astronomer did not know how to mount the meridian circle. He worked feverishly to set the instrument on its massive piers without telling anyone, planning to refute Armsby with an unexpected unveiling of the mounted telescope. But the work was slow, and he kept it secret.

Gould's delay in installing the telescopes and his refusal to answer questions were two of the main charges made against him by the trustees. Bache insisted that the trustees were not entitled to ask Gould any questions because they never paid him a salary, which would have allowed them to make demands upon his time. Again, the difference in perspective is crucial. The trustees believed that the January agreement named Gould as director on condition that the instruments be set up as soon as possible. Gould's refusal to answer questions about the progress of this work seemed unreasonable to them. Suspicion and curiosity increased unnecessarily with the secrecy.[8]

Gould was also greatly upset at this time by gossip circulating in Albany that he was a drunkard. Given Albany's fame as a center of reform—the state temperance society had its headquarters in the city—the local definition of drunkard was probably a stringent one. Although they were by no means intemperate, the members of the Lazzaroni fellowship certainly enjoyed a glass of hock or two when they gathered for their dinners. For his part Gould had a long-standing interest in good whiskey but being castigated as a drunkard was carrying it a bit far.[9] Gould bitterly resented such underhanded attacks through rumors and gossip. He considered these actions by his enemies to be unmanly and unworthy of gentlemen but vowed to stick it out, saying that if he broke down, it would be at his post. Martial images filled Bache and Gould's letters during this time, as each girded for the "great fight." Bache castigated the trustees as "a terrible set of dirty dogs."[10] Only Joseph Henry advised caution, suggesting that Bache tread carefully in assuming a partisan role in Albany. Henry was not uncritical of Gould either. In Henry's opinion Benjamin Gould was in an untenable position. The astronomer was both director of the observatory and a member of the Scientific Council to which he was appealing. Gould was, in effect, sitting in judgment on his own actions as director.

Although Joseph Henry believed Gould to be right in principle, the best interests of the observatory demanded a compromise acceptable to both sides. The scientists, Henry insisted, had to convince the public and the trustees as well that the good of the observatory was their first aim. Only insofar as this was accepted by the public, especially in Albany, were the scientists on safe ground. Henry did not believe that Gould's actions in Albany had been prudent. Joseph Henry had been born in Albany and knew the local situation well. He warned Bache that if Gould went on "to make battle on his own account, I fear he will put us into a very unpleasant position and damage his own cause." In a postscript Henry added, "Our might is in our right. Let it remain thus."[11]

Because of his own personal trials, Joseph Henry was sensitive to the vulnerability of institutions when bitter feelings were aroused in such controversies. He had come through two painful experiences himself—litigation with Morse over credit for the invention of the telegraph and congressional hearings over Henry's dismissal of Charles Coffin Jewett from the Smithsonian. Henry was aware of the inherent dangers for him at the Smithsonian and Bache at the Coast Survey in assuming a partisan position in the increasingly public hostilities over the Albany observatory.

Surprisingly, Bache—a consummate politician—allowed his temper and strong feelings to obscure his own normally acute political sense. Throughout the controversy Bache adamantly defended Gould against all criticism. As a result he appeared to be justifying arrogant and rude behavior, thus clouding the principles he was espousing. Bache, although respected as superintendent of the Coast Survey, was not without his own critics. By entering the Albany conflict as a hot partisan rather than as a cool arbiter, as Henry advised, Bache forfeited the credibility of a more detached position.

In arguing that the trustees of an antebellum institution had no right to interfere, even to ask questions about progress in setting up the instruments, Bache was far in advance of his time. Laymen could not be so summarily discounted in 1858. The first directors of scientific institutions were effective when they combined their scientific authority with persuasive skills, engendering enthusiasm for their institutional aims as well as their scientific goals. Bache himself was an excellent example of the benefits resulting from such a combination. In Albany it was just as imperative that Gould maintain the support of the trustees as it was to order the very best equipment for the observatory. It was the practical task of the director of a scientific institution to educate his patrons so that they could see the wisdom of his actions. Bache certainly understood the need to tend his own constituency, the Congress, carefully with regard to the needs and aims of the Coast Survey. His endorsement of Gould's antagonistic policies toward a smaller but equally important constituency of trustees at the observatory is thus more puzzling. Gould's actions were guaranteed to antagonize the trustees, who were, in an immediate sense, his clients.

Much of the literature on professionalization discusses the importance of the client relationship. In essence the client surrenders to professional authority because he lacks the necessary theoretical background to solve his problem. Unlike the lawyer or doctor, the scientist does not have an immediate client except, in an ultimate sense, society. For a scientist, the important judgments on his competency are made by his colleagues. As professionals, therefore, it is normal even today for scientists to have great sensitivity to their peers and less to laymen in positions of authority.[12] But institutional roles are more defined today. Charting a path that would safeguard both scientific competence and institutional growth was the task that faced emerging professionals like Benjamin Gould in antebellum America. The trustees were not attempting to usurp or challenge Gould's role as an astronomer. They were resisting his attempt to proclaim an institutional authority so total that it excluded them completely. Expert knowledge did not confer institutional autonomy in 1858. Joseph Henry had given good advice to the members of the Scientific Council. They had to appear to have "right" on their side before they could wield their might in Albany.

On 17 June, Olcott received the reply of the Scientific Council to the trustees' "want of harmony" resolution. Because the trustees had been intent on asserting their own authority after the May incidents, the Scientific Council's reply was provoking. The scientists insisted that they were the ones who held ultimate responsibility for the observatory, and therefore it was only appropriate that the trustees should appeal to them to resolve this dispute. A full statement of the difficulties "with such facts as may bear upon them," along with full copies of all the minutes of the trustees' meetings since 1 January 1858 was to be in the hands of the Scientific Council within three days.[13]

Power is a nebulous quality. Its appearance is often more important than its actual presence. If they accepted the role of appellant court, which the Scientific Council outlined for itself, the trustees would be reduced to a footing equal with the director whom they had dismissed. The implication was clear. General Robert Pruyn complained that he had gone along with the Scientific Council in the conflict over Peters, but he could not now support their claim "to be independent and superior to" the trustees of the observatory. The members of the Scientific Council were respected in Albany; but no one assumed they were impartial, especially as far as Gould was concerned. In a later statement, Bache would write that the Scientific Council had requested this information from the trustees as friends of the observatory, "not as partisans." But the scientists had attacked Peters too harshly and championed Gould's cause too ardently to be accepted as unbiased judges.[14]

Every week presented new problems. Tempers had been so roused that public attention was inevitably drawn to the quarrel. Albany newspapers began making inquiries about the reported "want of harmony" between the trustees and the director. In answer to the *Albany Atlas and Argus*'s request for more information, one trustee gave the newspaper a copy of John Wilder's angry speech at the trustees' June meeting. Wilder's speech, strongly condemning Gould and his assistants, was printed on

19 June. Gould immediately fired a salvo in reply. He had no wish to be drawn into a newspaper argument, he insisted, but could not let Wilder's accusations stand unanswered. Armsby and Wilder were guilty of rude conduct and foul language in their visits to the observatory and could not be excused. Of course, Wilder discharged a broadside in return. Like Gould, the trustee expressed his own unwillingness to enter a public controversy, but Gould had misrepresented the facts. Wilder concluded that he was not surprised Gould did not find his assistants' conduct rude, given the astronomer's own behavior.[15]

The pugnacious letters drew curious attention, especially in New York and Washington. The participants were familiarly known in both cities. The *New York Times* recounted the "belligerant correspondence" and noted that the conflict was "becoming decidedly personal." Gould immediately launched a volley of letters in protest to both the *New York Times* and the *New York Evening Post* for reprinting Wilder's charges against him along with his reply. No quarrel existed, Gould argued, because he would not enter a public controversy. He described the newspapers' publication of the articles as an attack on his personal and scientific character, suggesting that editors should refrain from publishing inflammatory statements when propriety precluded him from replying. Although he expressed his regret that the situation had arisen in Albany, Gould insisted that he had followed the only course compatible with his ideas of duty. As a professional astronomer, he argued, his entire life had been devoted to science, and only competent scientific tribunals, such as the Scientific Council, were capable of judging his actions. Only his professional colleagues could judge his competency. Public opinion had no relevance in the quarrel.[16]

Gould's close friend, Wolcott Gibbs, made plans to publish a notice asking for a suspension of public opinion until the members of the Scientific Council rendered their decision. Although the scientists intended to draw the curtain of professional authority around the Albany problem, the damage was already done. Gould's angry letters to the New York papers simply aroused more interest in the quarrel. Julius Hilgard, supervisor of the computing department of the Coast Survey, warned Bache that the controversy was attracting a lot of attention in Washington.[17]

As public interest widened, the scientists planned their next move. they decided to repeat the tactics that had successfully ousted Peters in January. The members of the Scientific Council would go to Albany and meet personally with the trustees. Bache believed that their combined prestige would so overwhelm the trustees' resistance that they would see that their dismissal of the director without the approval of the Scientific Council was presumptuous. Despite Joseph Henry's aversion to traveling in the heat, the scientists made plans to gather at the observatory, remaining there for three days while examining Gould's "progress in scientific matters, which we shall find to Gould's great credit, under these formidable circumstances." Peirce agreed, but Joseph Henry absolutely refused to leave Washington before the end of June. Bache continued to insist on Henry's presence, mindful of the importance of the physicist's reputation and relationships in Albany.[18]

As Bache, Peirce, and Joseph Henry worked out arrangements, matters were already moving toward a climax in Albany. The trustees convened on 26 June 1858; Olcott read the Scientific Council's letter aloud. The tone of the letter convinced nine of the trustees that the scientists were attempting to assume "an ultimate jurisdiction" over the observatory. If this were accepted, they were certain that Gould would be retained as director despite the wishes of the trustees. They inferred from the letter that the Scientific Council believed it could reverse the decisions of the board of trustees at any time.[19]

Olcott then read a lengthy statement of his own, recounting his many sources of dissatisfaction with Gould as director of the Dudley Observatory. This statement offers a revealing summary of the perceptions of the majority of the board of trustees, but is in its own way as self-serving as the later pamphlets. In addition to legitimate complaints about Gould, Olcott and other members of the board also wished to shield themselves from full public responsibility for the great amounts of money already expended in equipping a nonoperational observatory. As Gould had absolved himself in January from responsibility for expenditures, so now did Olcott hope to diminish public impressions of poor financial management by the trustees.

Olcott accused Gould of "gross mismanagement of the funds and general impracticability of character and conduct." The bank president blamed the astronomer for spending large amounts of money with little result. He made it clear that the trustees had been providing money to cover Gould's requests and expenditures out of their own pockets so that the endowment would not be encroached upon further. The trustees, said Olcott, had been persuaded by the array of gifted names on the Scientific Council to place unquestioning confidence in Gould's ability to bring the observatory into operation. They had complied with every request for funds that Gould had submitted until, at the end of 1857, it was increasingly apparent that Gould was reluctant to commit himself to the institution. As spokesman for the trustees, Olcott said he could no longer remain silent when Gould attempted to blame the board for expenditures made at the astronomer's own request. Olcott was equally incensed by Gould's comments, both in private conversation and public journals, that the astronomer was supporting the observatory out of his own pocket. In Olcott's words, Gould's "insinuations at home and abroad" that the observatory "was a failure and an abortion" destroyed public faith in the institution just when it was most needed. The bank president recounted the terms by which Gould was named director. Despite repeated explanations by the trustees that no money was available beyond a minimum amount necessary to set up the instruments already in storage, Gould continued to request additional funds, persisted in advancing his own money without approval, then demanded repayment from the trsutees for what he had spent.

Olcott related the history of the trustees' relationship with Dr. Gould from his own perspective, illustrated with numerous examples of Gould's "extravagances."[20] Obviously, the great turning point for Olcott with regard to Dr. Gould was the acrimonious trustees' meeting in January of 1858. Although the Scientific

Council's insistence on dismissing Dr. Peters had been perplexing, Gould's unexpected disclaimer of responsibility for expenditures had truly shocked the bank president. Nothing had been done at the observatory without Gould's full approval, "not even to the driving of a nail." Olcott argued that Gould's absolute repudiation of responsibility caused the trustees as a group to lose faith in the astronomer's integrity. Despite the acrimonious conflict over Peters, Gould's rejection of responsibility was the important issue for Thomas Olcott. Judgment on personal character was an important part of the bank president's everyday dealings. To put it simply, he no longer trusted Benjamin Gould.

As president of the board of trustees, Olcott argued that the one course of action that would quickly restore public confidence in the observatory—setting up the telescopes—had been repeatedly urged by the trustees and continually resisted by the director. If we are to believe Olcott, we have to infer that neither he nor any of the other trustees were aware of Gould's work on the meridian circle. Nor does Olcott's private correspondence indicate any knowledge of Gould's estimates of the time required to complete that installation. Given his habits of secrecy and his desire to confound his enemies, it is obvious that Gould sought to shame his critics with a spectacular and unsuspected unveiling of the telescope. This pattern of behavior unquestionably worsened an already difficult situation.[21]

Gould had systematically excluded the trustees from knowledge of the work going on at the observatory. As far as they could see, he was concentrating on installing a clock system that would not be needed until the heliometer arrived; meanwhile, the meridian circle and transit remained in their crates. In the absence of other information, Gould's seemingly illogical set of priorities was attributed by Olcott to the astronomer's "excessive particularity." That obsession with detail, again characteristic of the manic-depressive personality, had already been cause for comment. Gould had not explained his reasons for rejecting the first set of foundation stones for the telescopes. His insistence on constructing his own crane to move the massive stones into place rather than allow the masons to do the work was equally perplexing. His iron shutter "brainstorm" had proved disastrous for the dome. The director appeared to be requiring perfection in minor details that were doubly expensive and often ill-starred. Despite all the work and all the money spent, as far as the trustees knew, the observatory had made little progress. Why, asked the president of the board of trustees, did such magnificent telescopes sit in their crates for a year or more? Olcott answered his own question: "It is a discreet unwillingness to test his skill as a *practical* astronomer." This charge, more than any other, was most resented by Gould's supporters. The Scientific Council later responded that Olcott, as a layman, could not make such a judgment about a professional.[22]

Olcott also raised the sticky question of Gould's apparent inability to cooperate with other American astronomers. He referred pointedly to Gould's refusal to connect the Dudley Observatory with the Harvard Observatory during the longitude determination. Gould's dislike of William Cranch Bond and George Phillips Bond of the Harvard Observatory was well known. Thomas Olcott was blunt in his assessment. Gould's unwillingness to work with the astronomers at Harvard had increased the chances of error as well as the cost of the work of determining the Dudley Observatory's longitude. This was one of the strongest charges made against Gould by Olcott—that Gould's personality impeded the effort to establish the observatory as a nationally respected institution because he could not bring himself to work with men such as the Bonds, O.M. Mitchel, Matthew Maury, or Franz Brünnow. Olcott was clear about the cost of Gould's behavior and attitudes, describing arrogance in a director as an insurmountable obstacle for any institution in its early days. He emphasized that the Dudley Observatory had sizeable tangible assets and, with the right director, could take the place planned for it in the world of science.[23] But, he insisted, the future of the institution demanded a new director, someone other than Benjamin Apthorp Gould, Jr.

Leaving the most dramatic quarrel until last, Olcott argued that Gould's letter to the trustees praising the rude conduct of his assistants was so arrogant, insolent, and full of conceit that it alone was sufficient cause for terminating his relationship with the observatory. The trustees could not submit further to Gould's contempt or his taunts. In concluding his remarks, Olcott then gave his answer to the Scientific Council's demands. The scientists erred if they believed the trustees were appealing to them as a higher authority in this matter. The question of retaining Gould as director was no longer debatable. It should not even be open to further discussion.

Judge Ira Harris then rose to speak. The Scientific Council, he said, had obviously misunderstood the trustees' earlier "want of harmony" resolution. Harris offered a new resolution asserting that the "immediate withdrawal of Dr. Gould" from the observatory was absolutely necessary to the future of the institution. Emphasizing that the trustees did not recognize the right of the members of the Scientific Council to review the actions of the board, Harris argued that compliance with the scientists' request for information could be of no possible use. The question of retaining Gould was no longer open for consideration. All of Gould's relations with the observatory must be severed, including his membership in the Scientific Council.

After these two speeches, the meeting grew quite heated. Four trustees did not want to offend the other members of the Scientific Council. No one defended Gould's conduct, but a minority wished to be more cautious. The majority remained adamant. The Scientific Council had to understand that Gould's relations with the observatory must be completely severed. The majority were convinced that no facts would induce the members of the Scientific Council to consent to Gould's removal. As no arguments would convince the members of the majority to retain Gould as director, the issue was effectively reduced to one fundamental question: whose authority over the Dudley Observatory prevailed? In a vote of eight to four, the trustees passed Judge Harris's resolution, affirming their right to make an irrevocable decision on the question of who should be the director of the observatory. With the publication of the resolutions, the dispute between the trustees and the Scientific Council over Gould was obvious to anyone who could read a newspaper. The *New York Times* endorsed the trustees completely, but the *Albany Atlas and Argus* was more guarded, stressing the need for more facts. In the meantime, Bache, Peirce, and Henry arrived in Albany.[24]

On Tuesday, 30 June, the members of the Scientific Council sent off a letter to Olcott, requesting specific evidence to support his allegations against Gould. They argued that Dr. Gould had been selected as the most qualified director, and his rejection without sufficient cause would cast a public aspersion on the competency of the other members of the Scientific Council. They could not sanction his removal unless convinced he was unfit. If proof were not made available at once, they would assume that Gould was being subjected to unjust and cruel persecution by the trustees. Should Gould be removed capriciously, the scientists warned that no proper astronomer would take his place. They announced that the Scientific Council would be in session at the observatory on 1 July 1858 at nine in the morning. Olcott must present his evidence on or before that time. Olcott replied on the same day. While professing his profound respect for the individual members of the Scientific Council, he refused to recognize them as an appellant tribunal. He would not comply with their summons.[25]

The members of the Scientific Council then launched into print, issuing a letter and resolutions to the press. In the public eye, they said, they had full responsibility for the scientific conduct of the Dudley Observatory. The trustees had taken advantage of that public impression to secure donations. If the Scientific Council risked their own reputations, the trustees must "necessarily have delegated" all powers essential to the performance of those responsibilities by the scientists. The director was the Scientific Council's chosen agent at the observatory. Therefore, according to the scientists, he could not be dismissed by the trustees without the full approval of the Scientific Council. Olcott's list of charges against Dr. Gould should have been referred to them rather than to the board of trustees. The Scientific Council "should have at least concurrent, if not exclusive, jurisdiction." With this introduction, the council then made public its request for evidence from Olcott.[26]

The scientists' position was strengthened by a letter from Mrs. Dudley calling upon the majority of the trustees to resign. This letter implied that her donation to the endowment would be withdrawn unless the trustees cooperated fully with the members of the Scientific Council and retained Dr. Gould as director. By this time, of course, Mrs. Dudley's nephew had taken control of her estate. The Scientific Council published her letter, but local comment in Albany raised serious doubts as to whether she had really written it. Referring obliquely to those doubts, the trustees replied by saying that they were fully "aware of the influences which induced you to make the communication." They accused the man who dared to invade the seclusion of such an aged lady of "moral forgery" and "unworthy motives."[27]

As they gathered their forces to rebut Olcott's well-publicized allegations against Gould, the scientists hoped to close the ranks of the profession around him. Gould wrote George Bond, requesting a letter from the Harvard astronomer that would refute the "libelous statement" made by Olcott about Gould's refusal to cooperate with the Harvard Observatory. Bond replied that he had read Olcott's remarks in the *New York Times* for himself. The Harvard astronomer stated that he thought Olcott had been fairly close to the truth on the question of the longitude determination, saying that Olcott seemed to have drawn a perfectly natural inference from the facts. When Gould replied that the state legislature had required him to use a point within the boundaries of New York, Bond punctured that excuse by reminding Gould that Gould had suggested that restriction himself. The two astronomers exchanged several letters, but Bond stood by his position. He believed that it was only one among too many instances in which Gould had allowed "unfortunate personal relations" to prevent a friendly relationship between the two observatories.[28]

George Bond was impatient with Gould's insistence that he deny that a real breach existed if Gould planned to do nothing to bridge it. Public denials of private realities to suit Gould's convenience did not appeal to the Harvard astronomer. If the shoe fit, he was content to let Gould wear it, however it might pinch. Instead, Bond wrote Peters that he would rejoice "if the public generally would judge as the Trustees' have done, not so much by what hath been *said*, or attempted as by what hath been *done*." In an 1859 letter to another Dr. Peters (C. A. F. Peters, director of Altona Observatory), Bond stated privately that the course followed by Gould and his supporters in the Albany controversy "never had the least sympathy nor approval from me." Bond thought it foolish to treat astronomy "as a Myth that none but the initiated can comprehend in its simplest details."[29] George Bond's refusal to close ranks with the professionals against an attack by the "uninformed" confirmed opinions among the Lazzaroni fellowship that they were correct in excluding him from positions of leadership within the scientific community. The Lazzaroni still held strongly to behavior that Joseph Henry would later judge disastrous for American science—"supporting our friends right or wrong."[30]

During this fruitless exchange between Gould and Bond, the members of the Scientific Council made another effort to avoid a head-on conflict. They offered the trustees several options, hoping to use the weight of public opinion to force a compromise. They first proposed that the Scientific Council, as a group, take over the observatory completely. This was, of course, an effort to retain Gould as acting director while the other scientists, nominally in charge, returned to their regular positions. Secondly, the dispute might be referred to five "discreet and wise citizens of Albany." Two were to be named by the trustees, two by the Scientific Council, and these four would select a fifth. This had the advantage of placing the scientists on equal footing with the trustees in numbers within the committee and in public opinion. The third alternative suggested that a committee of three trustees be appointed, representing both the majority and minority on the board, to work out some compromise. This option would circumvent the active hostility of nine members of the majority, by reducing their numbers to two who might be more easily swayed by Bache, Peirce, and Henry. Olcott replied that the immediate dismissal of Gould was "the *only* basis of an amicable adjustment" in the problems besetting the observatory. The response from the scientists was equally adamant against dismissing Gould. On Saturday morning, 3 July 1858, the trustees met and severed relations with the Scientific Council.[31]

In the resolutions adopted at this meeting, the trustees insisted on their right to control the observatory through powers conferred on them by the state legislature in the observatory's charter. The charter empowered the trustees alone to appoint officers or servants. If they could appoint, they could dismiss. They had dismissed Dr. Gould for causes they thought sufficient. The scientists were attempting to assume "powers with which they were never clothed" and to exercise authority not legally theirs. The Scientific Council, they stated, had no right to hold the observatory against the will of the trustees or to set themselves up as managers of the institution if the trustees did not desire it. Faced with what they saw as a usurpation of their lawful authority, the trustees decided they had no choice except to sever relations completely with the council. Olcott was ordered to take whatever measures necessary to secure possession and control of the premises. The board then appointed O. M. Mitchel as director of the Dudley Observatory.

The appointment of Mitchel was understandable from the trustees' point of view. The observatory originated in enthusiasm generated by that astronomer's lectures in Albany. Mitchel had designed the building, and the first subscriptions had come in on the understanding that he would be the director. It was natural for the trustees to turn to him. Unfortunately, as far as posterity's judgments, Mitchel's reputation is that of a popularizer. The impressive standards of the future would be set by those who resembled Gould, not Mitchel. Although Mitchel was not educated as an astronomer like Gould, few antebellum practitioners were. Mitchel's florid style, which can perhaps be described as qualitative, was in sharp contrast to Gould's quantitative precision. Yet, in the 1850s both Mitchel and Gould concentrated on the same kind of astronomical work, compiling stellar zone catalogues. The contrast in personal style is not necessarily informative with regard to the meaning of Mitchel's appointment in this conflict.[32]

Mitchel's appointment at this juncture in the controversy offers a seductive simplification of the issues. It reduces a very complicated situation to the black/white terms of professional science versus popularization, elitism versus democratic values, the future versus the past. However appealing this reduction might be as an example of the problems of professionalization of American science, the simplification does not explain what happened in Albany. It cannot be assumed that the trustees were rejecting professionalism and embracing popular entertainment when they appointed Mitchel. What *can* be assumed is that the trustees were rejecting Benjamin Apthorp Gould, Jr., as director of the Dudley Observatory.

CHAPTER 10

The Pamphlet War Begins

Clearly determined to hold the observatory, Bache began strengthening his position. He wrote first to Howell Cobb, secretary of the treasury under whose jurisdiction the Coast Survey operated, to justify his continued presence in Albany. He warned the secretary that valuable Coast Survey instruments were in jeopardy and enclosed a letter from Congressman Erastus Corning that declared Bache's presence in Albany to be indispensable. Cobb authorized Bache to remain in Albany as long as necessary.[1]

In the meantime the trustees' resolutions were delivered to the observatory, along with a cover letter from Olcott giving the Scientific Council one month to leave the premises. Gould's staunch friend, Wolcott Gibbs, conferred with Theodore Sedgewick, district attorney for the Southern District of New York. Sedgewick felt he owed Bache a favor but cautioned that he could only give private advice. Albany was out of his district. He suggested Bache contact James C. Spencer, district attorney for the Northern District of New York, if he wished official action.

He also advised Bache to obtain written authorization from the Treasury Department that would enable Spencer to act at once.[2]

Bache immediately wrote Spencer, who was traveling the northern circuit, insisting that he come to Albany at once to protect government interests at risk at the observatory. Bache rested his right to maintain possession of the premises on the grounds that the Dudley Observatory was an *official* station of the Coast Survey. This was, of course, a complete reversal of Bache's earlier position. All through the spring and summer of 1858, Bache and Gould had meticulously denied any official connection between the survey and the Albany observatory. Bache now described Gould to Spencer not as the ousted director of the Dudley Observatory but as his official assistant in the Coast Survey. He warned that the federal government would suffer great losses if the survey were forced to give up the property because it would invalidate all previous work done in the Hudson River triangulation. In addition to these dangers, the delicate instruments, which were federal property, would undoubtedly suffer injury if his assistants were forced to leave.[3]

Spencer replied from Utica the next day. If the Coast Survey held the property by virtue of an agreement with the trustees, Bache would have to depend on the strength of that contract to maintain possession. If the arrangement was by virtue of the trustees' consent, then the Coast Survey was only a tenant at will or sufferance. In this case a month's notice to quit the premises was legally sufficient to force the survey out. This was disappointing news, but Spencer added an encouraging postscript. If Bache believed the Coast Survey had a legitimate right, it was his duty to remain in possession of the observatory. There might be a way around the law if "great results that would be lost in any other course" could be used to justify a willful trespass by the Coast Survey. Spencer limited his own responsibility by stressing that he did not know all the facts and could not act without official authority. He urged Bache to consult with the secretary and solicitor of the treasury as well with an Albany attorney. Fearing forceful expulsion from the observatory, Bache quickly consulted several Albany lawyers, including the respected John

H. Reynolds. Reynolds calmed the agitated superintendent, assuring him that it was unlikely that the trustees would force the scientists out since the eviction notice gave Bache possession until 7 August. Reynolds knew that most of the trustees left the city on vacation during August anyway, so there was no need to insist that the district attorney rush to Albany.[4]

Despite Reynolds's calming words, excitement ran high among the scientists' friends. John V. L. Pruyn, a respected member of the New York Board of Regents, returned from Newport in early July to find Albany in an uproar. He wrote in his diary that affairs at the observatory were "in a very unfortunate condition." Pruyn met with his relative, General Robert Pruyn, who complained lengthily of Gould's behavior. John Pruyn urged the general to adopt a conciliatory position in the conflict; General Pruyn promised to think this over, but the insults had struck deep. In the meantime friends of the Scientific Council were busily going over printers' galleys of the *Defence of Dr. Gould*, a pamphlet written by Bache, Henry, and Gould. In spite of his misgivings about Gould's position and conduct, Joseph Henry had assisted in the pamphlet's preparation.[5]

On 10 July 1858 the Scientific Council published a long letter in the Albany papers in which they expressed their astonishment that Thomas Olcott, a bank president, presumed to pronounce judgment on the question of Gould's competency, stressing the great esteem in which the astronomer was held by scientists throughout the world. The members of the Scientific Council then contrasted their own desire to create an institution that would advance knowledge with Olcott's desire for notoriety, comparing the staid and proper bank president to a charlatan setting off fireworks and sounding trumpets. The letter did little to ease hostilities. On the same day, the first copies of the *Defence of Dr. Gould* came off the presses. Gould warned Peirce to be careful about showing it around, as he hoped to surprise "the Enemy" by distributing it all at once in several cities. The first printing of fifteen hundred copies was gone within a week. Fifteen hundred more were ordered. Some copies escaped in Albany.[6]

General Pruyn, who had been moved by his conversation with John Pruyn to attempt a starring role as mediator in the quarrel, obtained a copy of the publication. Upon reading the *Defence,* he quickly informed his relative that the remarks it contained made any attempt at conciliation out of the question. In their pamphlet the members of the Scientific Council had abandoned the trappings of impartiality in favor of the lusty grapplings of hand-to-hand combat. The *Defence of Dr. Gould* was a direct and open challenge to the trustees' control of the observatory.[7]

The trustees were condemned on all counts. In thirty-two sections the scientists argued that Olcott's charges against Gould were frivolous and without evidence, motivated only by Olcott's determination to crush Gould completely. Olcott was accused of such flagrant mismanagement of the observatory's funds that the Scientific Council felt forced to call for a public investigation. Gould was praised for the "temperate and conciliatory character" of his conduct and for his generous devotion to the observatory. Olcott, in contrast, was castigated for his "suspicious and irritable temper"; Dr. Armsby, John Wilder, and General Pruyn were charged with rude behavior unworthy of gentlemen. The nine members who comprised the majority of the board of trustees were condemned as a group for their "constant and harassing interference." Gould's gentlemanly conduct was extolled, and his young assistants were completely exonerated. The pamphlet implied that those who criticized Gould were probably liars and perhaps cheats, but certainly not gentlemen.

The verbal extravagance of the scientists' pamphlet strengthened a suspicion for which groundwork was already in place. Were the scientists disinterestedly seeking the advancement of knowledge as they stated? Or were they attempting to increase their own power to control American scientific institutions? As lengthy as it was, the *Defence of Dr. Gould* offered no argument that justified the Scientific Council in holding property that was not their own. Many who had watched in frustration as the "Cambridge clique" gathered the strings of influence in the developing institutions of American science cast critical eyes over the explanations offered by the "self appointed guardians of science." Critics of the Coast Survey, of Lazzaroni influence in the American Association for the Advancement of Science, of Peirce's behavior in the American Academy—all cheered on the sidelines as the Scientific Council began to take some lumps in the newspapers. The weight of newspaper and magazine opinion came down against them. Only the *Boston Courier* supported the scientists; the *Boston Advertiser, Boston Journal, Evening Transcript, Daily Traveller, New York Times,* and the *Philadelphia American and Gazette* were critical.[8]

As copies of the pamphlet were shipped off across the nation to be used as ammunition by friends, Bache decided that the battle required his continued presence on the front lines. Instead of spending July in Maine, he would remain at "Fort Dudley," marshalling his forces to defeat the foe. Like the well-trained graduate of West Point that he was, Bache acted first to secure his line of supply. He directed his assistant in Washington, Captain William R. Palmer, to advise the solicitor of the treasury, Junius Hillyer, that the trustees were violating an agreement by trying to evict the Coast Survey before observations were completed. The solicitor should authorize the district attorney to support Bache fully in the conflict. In the meantime Bache wrote Spencer to arrange a meeting, promising that he would bring authorizations from the Treasury Department.[9]

As Bache worked to bring the massive artillery of the federal legal structure to fortify his position at the observatory, he and Henry outflanked the enemy in their own territory on 13 July. A public meeting was called by "those disapproving of the course of the majority of the trustees." Despite a violent storm, the crowd was large and enthusiastic. Bache spoke for over two hours, frequently interrupted by applause. A committee of five was appointed to draft resolutions expressing public disapproval of the trustees' actions. The resolutions censured the trustees and insisted that the observatory should remain in the hands of the scientists. Furthermore, any astronomer who replaced Gould in such circumstances would betray his brotherhood in science. The meeting lasted over three hours, but few left until it finished. Bache sent copies of the resolutions off to the Associated Press and described the evening to Peirce as a splendid one. He lightheartedly reported that the trustees were in a tight place on their own home ground. A committee of fifteen respected citizens—including Erastus Corning, Harmon Pumpelly, several other business and political rivals of Thomas Olcott, and friends of Joseph Henry and Benjamin Peirce—was appointed to prepare a more lengthy statement. This was published as a pamphlet in August.[10]

This pamphlet emphasized that the observatory belonged to science, not to Albany or to the trustees, and offered a very circumscribed interpretation of the trustees' institutional role. According to the authors, the trustees were simply there to fulfill the legalities of the state charter and were only incidentally concerned with the institution. They could have no part in bringing the observatory into operation or, more importantly, in judging Gould's competency as director. The Scientific Council, on the other hand, had unlimited authority. The writers of the pamphlet offered a new interpretation of institutional roles for antebellum America. A scientific institution, they argued, was a distinct entity that could not be conducted like any other. Nor could the scientists be held to traditional standards of accountability.

> For such a board of Trustees to assume the arbitrary right of appointment and removal over the scientific Director of an Observatory, according to their own unaided judgment concerning his capacity and fitness, or according to the fancies they may indulge, or to assume the right of professional supervision, direction and control over him in the performance of his scientific duties, just as they would exercise the same powers over a paid clerk or agent, if they were the Trustees of a Railroad or a Transportation Company, would be to mistake very grossly the nature of their own proper duties as well as the nature of his.[11]

This was a sweeping assertion of professional authority in 1858. According to the writers of the pamphlet, the trustees were not even empowered to dismiss the director of the institution without the unanimous approval of the scientists. Furthermore, they could not dismiss the Scientific Council at all, for

"Science was their master—not the Trustees." The pamphlet does not explain why it was possible for the trustees to engage the Scientific Council but not to dismiss them, a distinction important in antebellum America. This model of professional authority extended far beyond normal expectations in either professional or philanthropic institutions. Although a professional is one who is sought out so that he might provide direction and advice, lawyers and doctors were certainly dismissed at will, as were directors of orphanages and hospitals. Scientists in other positions, those at the Coast Survey included, did not possess this kind of security.

The pamphlet raises interesting questions with regard to informal understandings about the employment of scientists before formal tenure came into being. By 1858 the Dudley Observatory was not associated with an existing university, but the Scientific Council's rejection of the trustees' right to dismiss Gould has the overtones of an academic dispute. The only institutional situation that offers a comparison to the Scientific Council's claim that Gould could not be dismissed without the consent of his professional peers would be in the colleges, where these questions were by no means settled at mid-century. Faculty and trustees in antebellum colleges were still grappling with sticky disputes over the parameters of faculty authority. As Jasper Adams pointed out in 1837, it was not unusual for the faculty of a college to be described as "the servants of the trustees."[12] The Harvard charter specified that the acts of the corporation were valid only after the consent of the overseers had been obtained. The overseers elected the presidents, paid the tutors, and set the rules for "the college servants." William and Mary came the closest to self-government, in that the corporation of the college, consisting of the president and the masters, had the financial management of the institution in their own hands. At Yale and Princeton, there was no question of self-government. As Richard Hofstadter points out, the American teaching profession has traditionally been subject to government by an oligarchy of laymen. As much as the faculty tried to change these procedures, final authority remained with the trustees. Although antebellum faculties were given wide latitude in function, sweeping assertions of faculty authority were rejected at Harvard and other colleges. Not only did ultimate authority rest with antebellum governing boards, but in most twentieth-century academic institutions, according to Richard Hofstadter, "crucial decisions on university policy are [still] made without consulting the faculty."[13]

In addition to its questionable justification with regard to antebellum professional-client relationships or academic practices, the severely limited version of the trustees' authority offered in this pamphlet and the Scientific Council's own publications rested on shaky legal grounds. The writers acknowledged this, saying that the charter might very well give the trustees the legal right to turn Dr. Gould out into the street on a moment's notice, might even enable the trustees to lay violent hands on all the delicate instruments, or commit other "acts of positive barbarism." They did not intend to discuss the matter in terms of legalities: "We shall speak only of obligations, such as faith, and truth, and honor, and conscience make binding on every man." The pamphlet accused the trustees of failing to fulfill the "higher moral obligations of right and duty" with which they were entrusted.[14]

The trustees refused to respond. As each side watched the other, news from Washington indicated that the logistics of Bache's battle plan were encountering difficulty. Howell Cobb, secretary of the treasury, was absent from Washington. The acting secretary, Mr. Clayton, refused to authorize the district attorney to provide legal counsel to Bache. Clayton could see no connection whatsoever between the treasury and the Albany observatory; he refused to commit the department to legal action without further information. In the meantime the district attorney met with Bache. Gould described the results as encouraging, but Bache realized his legal artillery was bogged down.[15]

Although the trustees had not responded to the publication of the *Defence of Dr. Gould* nor to the provocation offered by the resolutions of the public meeting, rumors were circulating that they were preparing a statement of their own. John Wilder was said to be in charge of the task. On 15 July, two days after the tumultuous meeting endorsing the Scientific Council, Wilder spent the day closeted with Dr. Armsby, Judge Harris, and Thomas Olcott. Together, the four trustees drafted the *Statement of the Trustees,* as well as several articles to be given to the local papers. Later that afternoon John Wilder was stricken with chest pains. He returned with Dr. Armsby to his room at the Delavan Hotel where he died of "apoplexy." The trustees' pamphlet would be delayed.[16]

The secretary of the treasury returned to Washington, and Bache quickly requested permission to engage the district attorney in behalf of the Coast Survey's interests in Albany. Cobb gave guarded permission, authorizing Bache to seek legal counsel but warning him to be careful of the cost. Bache immediately advised Spencer that the course was clear for the district attorney to act officially. Glossing over the restrictions imposed by Cobb, Bache wrote that he hoped the district attorney would accept the conditions "with *philosophy,*" and not let the terms of Cobb's letter interfere with his actions in any way. Bache enclosed Olcott's eviction notice and asked Spencer if his own presence at the observatory was necessary to maintain the Coast Survey's authority over the property. He longed to join his friends at the Coast Survey station in Maine. As July slogged to its sweltering close, Peirce believed that any pamphlet the trustees could produce would benefit the Scientific Council, and Gould's friends were confident of victory.[17]

Observatory affairs remained quiet. Gould was suffering what he described as "the constant and mischievous efforts of designing men" to affect his mind by making his friends suspect him. He took the opportunity to leave "Fort Dudley" for a brief respite from the heat, traveling to Sharon Springs to soothe his headaches. He wrote John LeConte, a Lazzaroni friend, that the situation remained unchanged—"a condition of armed neutrality."[18] Both Gould and Bache returned to Albany in the second week of August to confer with Spencer, then Bache left for Maine.

Bache had consciously avoided giving Spencer specific instructions, hoping that the district attorney would act on his own initiative. The Coast Survey would be open to direct criticism if Bache requested any action himself. Secretary Cobb had only authorized Bache to secure minimal counsel from Spencer. He had not authorized any other action. Unfortunately for Bache, Spencer was a careful man. He wrote Bache several times, hoping to receive orders upon which he might officially act, but received no replies. Perplexed, Spencer did the one thing Bache wanted to avoid. Spencer wrote to the Treasury Department himself. He also enclosed his own five hundred dollar estimate of the cost for legal services provided by the district attorney's office in Bache's behalf so far.

Bache derived power from Spencer's involvement in the Dudley Observatory controversy only if the district attorney assumed a broad interpretation of the authorization he had received from the secretary of the treasury. If the instructions were filtered through Bache to Spencer, Bache could blur the restrictive edges of Cobb's authorization so that the district attorney would assume greater latitude. Bache believed that official support from the district attorney would work to diminish the legal advantage enjoyed by the trustees in the public mind from their authority under the state charter. If the district attorney began to communicate directly with the Treasury Department, however, the "Coast Survey necessity" that the Scientific Council was using to maintain possession of the observatory would be more closely scrutinized. Bache knew that his superiors in the Treasury Department had no wish either to assume legal expenses for members of the Scientific Council or to avenge the insult to Gould. The all-for-one fellowship of the Lazzaroni made no difference to the close-pocketed men at the treasury. Direct communication between the district attorney and the Treasury Department would result in restrictions difficult for Bache to expand to the Scientific Council's benefit. The district attorney's letter was channeled to Junius Hillyer, the solicitor of the treasury. Hillyer conferred with Secretary Cobb. Howell Cobb had been a powerful Democratic congressman from Georgia and now held a cabinet post in Buchanan's troubled administration.[19]

Because political influence is so much a part of this story, it is important to recall the larger background in which the conflict over the Dudley Observatory was taking place. Political rhetoric in the decade leading to the Civil War was focused on sovereignty, in reality, a question of power and ultimate authority. Although it is impossible to summarize the causes of the Civil War in a few paragraphs, it is obvious that the antebellum political system depended on the ability of the political parties to avoid sectional polarization. Unfortunately, new territory acquired in the Mexican War precipitated a decade of bitter sectional jealousies. Traditional cross-sectional alliances and older systems of political reciprocity, which crossed sectional lines to the national benefit, were collapsing under the strain. By the 1850s Congress had become a public forum for conflicting visions of American government, conflicting opinions on the nature of the Union. Which came first, asked John Calhoun, the nation or the state? Like medieval theologians politicians in both North and South split logical hairs debating whether sovereignty rested in the states or in the federal government. This esoteric debate had serious practical consequences for federal power, especially with regard to slavery, an issue that joined morality to the legal arguments.

Most Americans had little interest in slavery; they were more intent on harvesting the benefits of economic growth. But passions engendered by the extension of slavery into the territories dissolved previous understandings and restraints. The question came to a head in the chaos of Kansas. In 1857, despite the fact that the overwhelming majority of Kansas settlers opposed slavery, the Buchanan administration was committed to seeing Kansas enter the Union as a slave state. During 1858, in the aftermath of the Dred Scott decision and Kansas violence, Congress was an arena in which a vitriolic sectional struggle was fought daily on the question of federal authority. The summer resounded with oratory—Lincoln's "House Divided" speech, the seven Lincoln-Douglas debates, and Seward's "irrepressible conflict" speech—all of which explored the underlying constitutional principles of the Union.

Given this background it is easy to understand why a Southern Democrat like Howell Cobb would not be an eager participant in an unauthorized seizure of private property by a federal agency, a seizure against the will of those legally entrusted by a state legislature. Secretary Cobb ordered Hillyer to obtain a complete report from Bache as soon as possible. Bache must answer certain probing questions that bothered the secretary. What was the "nature, character and extent of the rights" the federal government had in this observatory? "How, by whom, and from whom were they acquired?" Did Bache have a legal document to support his claims? Cobb also demanded precise information about the extent of damage the government would actually sustain if the site were given up. Bache was ordered to ascertain the costs of defending the government's interests in court as well. These were hard, specific, and uncomfortable questions.[20]

Bache knew he could not produce any written agreement to substantiate his claim that the Coast Survey had undisputed right to the observatory. During the previous year, he had been meticulous in denying any official connection between the Coast Survey and the Dudley Observatory. How could he now provide a legal basis granting control of the property to the Coast Survey against the wishes of the trustees? The solicitor demanded "the fullest and clearest information" from Bache at once. Hillyer's letter gave warning that the Treasury Department was not going to enter the Albany conflict unsuspectingly to bolster the position of the Scientific Council. When he received Hillyer's authoritative request, Bache sat down and wrote Spencer an irritated letter of his own. He reminded the district attorney that his authorization had come from Bache, not the Treasury Department. There had been no need, Bache said, for Spencer to communicate directly with either the secretary or the solicitor of the treasury. At the end of August, the district attorney still remained perplexed, protesting that he had not yet received any specific instructions from the superintendent of the Coast Survey. Bache had clearly failed to bring the powerful legal reserves of the district attorney's office into action against the trustees.[21]

The scientists seemed to feel, at this point, that the next move was up to the trustees. The arguments presented in the *Defence of Dr. Gould* were as yet unanswered, but some of the scientists' friends had doubts about Bache's course. Senator James A. Pearce of Maryland, one of Bache's close friends, cautioned that Gould would gain little advantage if he vanquished the trustees "at law and in public opinion" if he planned to stay in Albany. What good would it do Gould to remain as director if the trustees upon whom he would depend were bitterly opposed to him? As a practical politician, Senator Pearce could smell a Pyrrhic victory in the making.[22]

The peace imposed by summer vacations ended on Monday morning, 30 August 1858, when *The Statement of the Trustees* appeared. The pamphlet was thought to be the work of Judge Ira Harris, but General Pruyn, Thomas Olcott, and Dr. Armsby also participated in its preparation. Despite the death of John Wilder, the pamphlet showed the trustees had used the month of August to their advantage. Judge Harris had put together a skillful and persuasive document. Gould received a copy and hurried to New York to confer with Bache, who was there for the celebration honoring the first transatlantic cable. Both Bache and Gould were disconcerted, for the trsutees' pamphlet was far better than they had expected. Reports of the positive response it received in Albany were dismaying. Even more alarming, copies had immediately appeared on the desk of every member of Congress. The trustees were carrying the battle onto the home ground of Bache and Joseph Henry. Over twenty thousand copies of the trustees' *Statement* were widely circulated in the autumn of 1858.[23]

The pamphlet demonstrated Judge Harris's political skill. By directing the argument against Gould rather than the Scientific Council as a whole, the trustees narrowed the grounds of their quarrel to their legal right to dismiss a director who caused problems at an institution for which they were responsible. By attacking Gould, Judge Harris neatly circumvented a direct confrontation with influential supporters of Bache and Joseph Henry. The *Statement of the Trustees* turned Gould's arrogance directly against him by publishing excerpts from his own letters during the past few years. Gould's derisive comments about other scientists and public figures placed him in an embarrassing position with many of his friends and confirmed his enemies in their impressions of his contemptuous attitudes. The letters were adroitly emphasized with italics and small deletions to show Gould in the worst possible light, but the words were his own.

Gould considered this action to be final proof of the trustees' unmanly and ungentlemanly characters. The letters led to uncomfortable situations with many of his fellow scientists, including Benjamin Peirce. Gould warned that the trustees were attempting to destroy all support for him by embroiling him with his friends. The publication of "strictly confidential" letters prompted Gould to compile another pamphlet of his own correspondence in response. In the meantime both Bache and Gould pored over the *Statement of the Trustees*, filling many handwritten pages as they noted points of minor disagreement. The Bache papers at the Library of Congress include long notes in Bache's own handwriting. Unfortunately, the combination of Bache's irritation and his nearly illegible handwriting make many of these practically indecipherable.[24]

Preoccupied as he was with his own problems, even Joseph Henry agreed that the trustees' pamphlet had been skillfully prepared. The *Statement of the Trustees* was making a strong public impression. It would require a careful answer. As the strength of the trustees' pamphlet became apparent, Bache observed that the trustees had gained by the August delay and had done "the false side well."[25] A rebuttal would have to be equally well done.

CHAPTER 11

War on Two Fronts

As the trustees returned from summer vacations, early reports indicated that their pamphlet, the *Statement of the Trustees*, had been well received. Many, like the editor of the *Boston Advertiser* who had earlier supported the Scientific Council, were now openly critical, believing Gould to be guilty of a "want of good taste" in persisting where his services were no longer desired.[1] With this success the trustees attacked Bache's base of power. On 2 September 1858, Olcott wrote to the secretary of the Treasury.

Olcott explained to Howell Cobb that Bache had taken illegal possession of the Dudley Observatory "under the pretence of some authority connected with his official position" as superintendent of the Coast Survey. The trustees asked the secretary to relieve them from an unwarrantable interference in the operation of an institution for which they were legally responsible. The bank president's letter was supported by a statement from a respected Albany law firm that Bache's actions were completely unauthorized and illegal. Cobb handed the two letters to B. F.

Pleasants, who was acting solicitor of the Treasury during Hillyer's absence from Washington. Pleasants was nervous about taking action. After conferring with Bache's aide, Captain William Palmer, he decided to let the decision await Hillyer's return. Palmer telegraphed the news to Bache, who was visiting with Peirce in Cambridge.[2]

During July and August, Bache had hoped to mobilize federal authority in support of the Scientific Council by encouraging the district attorney to act on his own initiative. With the news that Olcott was mounting a serious flank attack through his Treasury Department superiors, Bache changed tactics. If the legal wheels of the treasury began to move on their own, the interests of the Scientific Council would not enter the decision-making process. Rather than urging immediate federal action, Bache now emphasized the danger of making any sudden moves in Albany.

Bache turned to his influential friend, Congressman Erastus Corning. Corning was not powerful within the Congress, but his great political and economic influence in New York made him a person to reckon with. Corning was to make it clear to the secretary of the treasury that Bache was acting under Corning's personal advice, warning that precipitous action would place the congressman at a personal disadvantage in his home district. If Secretary Cobb took any steps in the Albany dispute, he should be aware that he could involve himself personally as well as his department in political problems. On the other hand, Corning assured the secretary that if Cobb took no action, he would avoid all difficulties. Bache then telegraphed the faithful Captain Palmer. Palmer was to stress to the treasury officials that Bache had carefully guarded the department's interests, and the department should not interfere in a delicate situation.[3]

As the trustees had attacked the Scientific Council's position through Bache's Washington superiors, so now the scientists worked to parry the stroke through the state government of New York. Because the observatory had been originally chartered as part of the proposed University of Albany, it fell under the loose jurisdiction of the regents of the state of New York. The longitude grant made to the observatory had also been put under the regents' supervision. In order to use this authority as a counterweight against the trustees' charter, Gould wrote to J. V. L. Pruyn. He accused the trustees of interfering with the longitude determination authorized by the state legislature and placed under the authority of the regents. Gould urged Pruyn, as a member of the regents, to take action. The regents should protect the interests of the state by prohibiting the trustees from interfering with the Scientific Council.[4]

In the meantime Bache was traveling north to the Coast Survey station at Humpback Mountain in Maine. He telegraphed Palmer from Bangor, emphasizing the importance of the Washington end of the dispute. Palmer must "not let it go wrong" in Washington. Bache's staff arranged a meeting with Solicitor Hillyer, who had returned to the capital. Hillyer was immediately immersed in the problems raised by Olcott's letter to Secretary Cobb. Bobbing about in complex political currents, Hillyer had no information to steer by. On 10 August, he wrote Bache asking for a complete report on the facts of the conflict. He received no answer. Grasping at any straw, Hillyer then wrote to James C. Spencer, the district attorney. In his July letter to the treasury, Spencer said that he had fully investigated the situation and concluded that the Coast Survey had a legal right to maintain possession of the observatory until its observations were ended. Hillyer now asked the district attorney to send him a report containing the facts upon which his conclusions were founded. The solicitor alluded to Bache's silence, observing that he had hoped to receive some specific information from the superintendent of the Coast Survey by this time.[5]

Bache, of course, had no desire to have the district attorney and solicitor of the treasury communicating directly. Without information, the Treasury Department could do little to affect the scientists' position in Albany. No information meant that no restrictions would be imposed from above. Lacking explicit instructions, the District Attorney might still be persuaded to bolster the council's legal position with a favorable opinion reassuring the public about the legality of the Scientific Council's actions. The burden would then be on the trustees. On the other hand, Bache's freedom to maneuver would be seriously curtailed if the cooly appraising eyes of his superiors in the Treasury Department weighed the financial and political costs of the conflict.

The Albany dispute was fraught with political risks for Solicitor Hillyer, who was frustrated by his inability to elicit any information from Bache. How could he guide the department through such a perplexing set of conflicting interests? Both Olcott and Corning were influential men. While he waited for a reply from Spencer, Hillyer was not disposed to make waves. He assured Palmer he would take no action until he met with Congressman Corning and Secretary Cobb. Bache had to be content with that. He had delayed Olcott's thrust through Washington, but he telegraphed Palmer that the situation must be watched carefully. If it could go wrong anywhere, Bache feared it would be in Washington. By 13 September Corning and Joseph Henry had reached an agreement with the Treasury Department. No action would be taken until Bache made a full report to Hillyer. This was good news. Because he controlled the channels of information upon which action would depend, Bache believed he could guide the department's entry into the dispute.[6]

Meanwhile, back at "Fort Dudley," the Coast Survey crew continued their observations. Despite Gould's wishes to confound his critics with rapid but secret progress in setting up the instruments, there were as yet no permanently mounted telescopes at the observatory. Only the small comet-seeker used by Peters to discover the "Olcott Comet" was in use. That small instrument again proved its worth. On the night of 10 September 1858, George Searle, one of Gould's assistants, discovered an asteroid. In a stroke reminiscent of Dr. Peters, Searle invited Mrs. Dudley to name the newly found heavenly body. Although it is unclear who made the selection, the name chosen was truly appropriate to the embattled condition of the Dudley Observatory. The small planet was christened "Pandora." Searle shared the 1859 Lalande Prize of the Paris Academy of Sciences for the discovery. This time Gould did not insist that the mythological name be withdrawn in favor of a numerical listing. Instead, he seized the opportunity to write the Astronomer Royal in Greenwich to explain the publication of Gould's deprecating comments in his letters. Friends had told Gould that the pernicious Peters had written several European astronomers about his persecution in Albany. Gould offered his own version. Describing himself as "exhausted and almost overwhelmed," Gould blamed the observatory's problems on the "profligate squandering of the finances" by the trustees and the intrigues of Dr. Peters. He told Airy that he had been supported by all true scientists, with unimportant exceptions.[7]

As Gould dispatched his remarks to Greenwich, a disturbing letter was coming north to Bache from Washington. The departmental wheels put in motion by Olcott's request had been slowed by Erastus Corning, but they had not been stopped. Olcott's prickly letter remained unanswered. As legal counsel to the treasury, Hillyer could not avoid a reply; nor was Olcott a man to be disregarded. Olcott's accusation that the Coast Survey had taken illegal possession of the observatory could not be ignored. Given the animosities aroused by the Dred Scott decision and the heated debates in Congress in 1858 over the extent of federal authority, the last thing that the solicitor of the treasury wanted was to be involved in a lawsuit over the seizure of private property by a federal agency. As long as slaves were possessions, federal seizure of any private property was part of a much larger issue, one that ignited inflammatory rhetoric instantaneously. Hillyer was also hesitant to jump into local rivalries in Albany. Corning was a Democrat; Olcott was a rising power in the new ranks of the Republican party. The opposition of these two men on any issue involved a history of thirty years of business rivalries and political alliances. In addition to Congressman Corning, both Joseph Henry and Bache had strong ties with influential people in Albany. With the direct participation of Corning and Olcott, Joseph Henry and Bache in the conflict, the solicitor knew he was moving through a complicated situation involving Albany business rivalries, New York political conflicts, federal-state tensions, and national scientific ambitions.

Hillyer's previous requests to Bache for information had been ignored. A full month had passed. By 13 September, Hillyer inferred from private conversations with Bache's friends that the superintendent of the Coast Survey had no intention of replying. Hillyer had agreed that he would take no action until he received Bache's report, but he refused to remain indefinitely without information. The next day he demanded the facts needed to answer Olcott's letter. He reminded Bache that the 10 August request for information had come from the secretary himself. If Hillyer were to reply to Olcott officially on behalf of the treasury, Bache had to supply answers. Was there a written agreement? What were its terms? With whom had it been made? Although Hillyer avoided confronting Bache directly, it is obvious that he was highly irritated by the superintendent's silence. Bache responded quickly with a short but formal note in which he attributed the delay to a misunderstanding with his Washington staff. He promised to start the report at once and closed his message by emphatically denying that the Coast Survey had taken illegal possession of the observatory, as alleged by Olcott.[8]

While these letters were passing in the mail, Hillyer's request for information from the district attorney reached Albany. Spencer met with John H. Reynolds, the Albany attorney who had been advising Bache. Both agreed that the survey had a legal right to maintain possession of the observatory despite the eviction notice from the trustees. In their combined opinion, the survey could not be forced out until the superintendent decided that all the observations needed in the Hudson River triangulations were completed. The Dudley Observatory would be under Bache's legal control, through the authority of the Coast Survey, until he decided he no longer desired it.[9] This was good news for the scientists, but all depended, of course, on preventing the Treasury Department from ordering the Coast Survey out of the observatory. The Washington end of the conflict was still critical.

The Albany end of the dispute continued to favor the members of the Scientific Council. As Spencer and Reynolds endorsed Bache's right to hold the observatory, relations between the trustees and Mrs. Dudley and her nephew, Rutger B. Miller, had swiftly deteriorated. By this time Miller was in complete charge of his aunt's affairs. A letter written in early September, signed by Mrs. Dudley and Miller, requested a conference with the trustees. Olcott addressed the trustees' reply to Mrs. Dudley alone. The state of local hostilities can be seen in this small incident. Miller took this as a snub from Olcott. On Friday, 17 September, Olcott received a letter signed by Mrs. Dudley, sharply ordering him to address all his future correspondence directly to her nephew. Mrs. Dudley referred to the "deplorable condition" of the observatory and urged that the institution be placed under the control of the regents of the state of New York, an argument that followed the lines of action Gould advanced to J. V. L. Pruyn earlier. In addition the endowment should be taken from Olcott's hands and placed under the direction of Erastus Corning. A general appeal should then be made to the nation to secure the complete sum required by the Scientific Council. In an important deviation from previous understandings, Mrs. Dudley's letter stipulated that the entire endowment must be obtained before any individual contribution could be deemed obligatory. This was a modification—of doubtful legality—of the original terms of her gift, probably attributable to Miller's control of her estate. Unless these terms were agreed to by the trustees, her letter stated that the Dudley contribution to the observatory would be revoked.

Mrs. Dudley's letter was given to the newspapers by Miller on the same day. In discussing his aunt's demands, Miller added final flourishes to the ultimatum. Unless the trustees agreed immediately to all the terms, Mrs. Dudley would complain to the attorney general of New York that the trustees had not only proven themselves unfit but had also violated the charter granted by the state legislature. Miller accused the trustees of squandering the money under their control and mismanaging the observatory. In an adroit turnabout, he blamed the trustees for the long delay in bringing the observatory into operating condition. The charter granted by the legislature stipulated that an observatory must be brought into existence; Miller charged that the trustees failed to comply with this stipulation. The observatory was obviously not functioning, therefore the trustees should be removed. To complete the attack, Miller ended by insisting that the trustees should be held personally liable for all the money wasted through their mismanagement. Miller's statement was an open attack on Thomas Olcott. Demanding that the endowment be taken from Olcott and handed over to Erastus Corning, Olcott's bitter enemy, was an obvious insult. The affront from Miller was complete when he lodged a complaint against the trustees with the attorney general of New York on behalf of his aunt Dudley.[10]

To add to the trustees' troubles, Gould accused them of maliciously perverting his letters. The astronomer referred readers of the *Albany Atlas and Argus* to his own pamphlet circulating in the city at this time. Stung by Gould's assertions, the trustees promptly answered him with a heated denial. Deletions in his letters were attributed to the need for brevity; any alterations were accidental and never changed the meaning. The nine trustees closed by stating that they had the originals and would, if necessary to prove their honesty, publish them in their entirety, no matter how embarrassing that might be to the astronomer.[11]

Despite the aggravations Olcott remained purposeful as strategist for the trustees. The trustees had failed to persuade Bache's superiors to order the superintendent out of the observatory, but their Washington efforts had not been without effect. Olcott continued to concentrate on Bache's vulnerabilities. Scientists who supported the trustees' position were asked if they had any information critical of the Coast Survey, the Nautical Almanac, the Light House Board, or the Smithsonian. Olcott clearly agreed with Bache's assessment that if it could go wrong for the scientists, it would do so most quickly in the national capital. For that reason the *Statement of the Trustees* was widely circulated in Washington. By mid-September Joseph Henry's assistant at the Smithsonian reported that almost every person of importance in the city had received a copy.

Joseph Henry was aware that support for the trustees' position came from some who resented the authoritarian attitudes of many of Bache's friends as they gained influence in American scientific associations and institutions. To many, Gould personified that arrogance. To these often discontended observers, the Dudley Observatory conflict presented a rare opportunity to see the haughty few who controlled many of the avenues for recognition in science receive what was regarded as an overdue comeuppance. Henry was also aware of strong hostility to Gould. Although he worried about the effect of the Albany conflict on the young astronomer, Henry did not allow this concern to obscure his own dispassionate analysis. Joseph Henry believed Gould's personality worked against the Scientific Council's position. This had been recognized and used by Judge Harris when he drafted the *Statement of the Trustees*; it was one of the factors that made that pamphlet so effective. The trustees' charges of misconduct were directed at Gould, not at the eminent scientists behind him. Given this emphasis Gould's personality and his erratic behavior became a critical issue. By centering the attack on Gould's conduct, the trustees successfully diluted the strength of the other scientists' reputations.

In such heated arguments, Henry knew that influences besides high principles weighed in the outcome. Gould's personality was, for Henry, an element that could sway the balance in the trustees' favor. As Henry observed to Bache, it was unfortunately true that Gould had "more personal enemies than any person with whom I am acquainted." These enemies would silently side with the trustees. Joseph Henry knew from bitter experience how subsurface resentments, springing from personal animosities but clothed in principle, could erupt in a conflict. His distressing problems with Morse and Jewett in the past years honed that awareness to a fine edge. Against this background of unpleasant conflicts, Joseph Henry could see that the trustees' pamphlet made a dangerously plausible and convincing argument. Although the hot-tempered Gibbs urged Bache to draft an immediate reply to the trustees, Henry cautioned that any rebuttal would have to be very well done and should not be attemped hastily. Henry warned Bache to expect "something of a fight" in Congress as a result of the *Statement of the Trustees*.[12]

While considering "Smithson's" letter, Bache set himself to the delicate task of placating Solicitor Hillyer. He drafted a skillful reply, adroitly skirting the specific questions asked in August.

He simply denied that the Coast Survey was illegally in possession of the premises. If that were true, Bache wrote, why didn't Olcott go to court to settle the question? If the Coast Survey was doing something illegal, why did Olcott appeal to Bache's superiors in the Treasury Department? By going to the treasury rather than the law for redress, Bache argued, Olcott was admitting he could not win in court. Therefore, the Coast Survey must have every legal right to remain at the Dudley Observatory. He then stated that Olcott had never asked him to remove the survey's equipment, flatly ignoring the pointed eviction notice presented in July. Without mentioning specific costs, Bache threatened grave financial loss to the government if the Treasury agreed to Olcott's request.

Bache lent weight to his argument by bringing forward the informal opinion of District Attorney Spencer. In a broadly sweeping assertion of federal power, Bache cited the enabling legislation by which Congress created the Coast Survey. The superintendent opined to the solicitor that, under the authority of this law, stations might be occupied on both public and private property. If the survey could be forced out of one important station, so might it be forced out of others. What held for one held for all. Bache also drew on the authority of the state government. The New York legislature had appropriated funds for a longitude determination. If the Treasury Department summarily ordered the Coast Survey out of the observatory without completing that work, the department would be liable to censure from the state legislature, from the regents, and from other prominent citizens of Albany, including J. V. L. Pruyn and Erastus Corning. Deftly circumventing the solicitor's own power within the department, Bache cited his authorization from the secretary himself, giving Bache authority to confer with Spencer. He did not refer to Cobb's restrictive references to costs in these consultations. Nor did he invite the solicitor's legal opinion on the right of the Coast Survey to occupy the observatory. Instead, he pointedly stated that he was content to leave such a decision to the district attorney, "under whose advice, by authority of the Secretary of the Treasury, I am now acting."[13]

The letter is fascinating. Every source of power—legal and political—at Bache's disposal has been brought into play against possible attacks arising from within the Treasury Department on his actions in Albany. In support, Bache offered Hillyer authority ranging from the secretary of the treasury, the district attorney, the laws of Congress, to the state legislature, the New York Board of Regents, and powerful local citizens. To balance this enormous weight of authority, Bache presented a list of dangers inherent in complying with Olcott's request. The government would suffer financial losses, political censure, and set dangerous precedents for future federal agencies doing field work.

But events in September were moving against Bache. The trustees' pamphlet continued to sap the strength of the scientists' arguments. Bache was in Maine, but Joseph Henry and the Washington staff of the Coast Survey knew that "the Chief" was paying an increasing price in Washington for his stand in Albany. The ever-loyal Palmer had written tentative cautions to Bache about his actions in Albany, but had been gently reprimanded. By the end of September, Captain Palmer was alarmed. He had been at the Smithsonian when a telegram arrived for Henry from Bache. Palmer described Henry as overwrought. Morse had made a new and devastating attack; "Smithson" decided that he could no longer afford to disperse his strength. Captain Palmer listened carefully as Henry insisted that

> . . . this whole "Scientific Council business" is getting to be a dirty business—which he has decided to *withdraw from* in as dignified and quiet a manner as he can.—He said he knows Gould will fly up at this, etc.,—but he must devote his energies to the Smithsonian duties, to science; to which he has devoted his life . . . he will go as far as any man for a friend, as far as he would for a Brother, but if when on the edge of a precipice, others will jump off, he won't, when it can do no good to any body.[14]

Palmer begged Henry to meet with Bache and Peirce in Boston, but the beseiged secretary of the Smithsonian refused, saying that he did not wish to "put himself among the Cambridge set now."

During this last week of September, Captain Palmer conferred with several other "leading people" in Washington. All agreed. Bache would be wise to withdraw immediately from all connection with the Albany observatory. Continued occupation of the property against the wishes of the trustees could only injure the Coast Survey. Palmer reported that one member of the President's cabinet warned that Bache should be giving his time and attention to his official duties rather than to private ventures at government expense in Albany. As superintendent, Bache would do well to look at the Albany situation "in a *practical* light" as far as the Coast Survey was concerned, Palmer advised. Anticipating objections, Palmer warned that members of the President's cabinet did not see the controversy from Bache's perspective. As far as they were concerned, the whole business was outside his proper scope of duties and laden with political liabilities. To persist in holding the observatory, Palmer feared, would result in so much abuse in the newspapers that all the institutions connected with the involved scientists—the Coast Survey, the Nautical Almanac, the Smithsonian, the Light House Board—would suffer the consequences.

Palmer then approached the delicate area that Henry had skirted. He spoke plainly. Everyone recognized Dr. Gould's ability, but Captain Palmer insisted that most who knew him would agree that Dr. Gould was extremely "arrogant and conceited."

> I [Palmer] know this is to a certain extent true.—from my once slight knowledge of Dr. Gould.—I venture the assertion confidentially to you—from my own limited knowledge, that a part of the Dudley troubles have arisen from the fact that Dr. Gould has wounded the vanity, pride, self-love . . . of some of these . . . trustees—I am confident he has been wanting in that smooth mode or manner, and apparent deference to them—which is so necessary. . . .[15]

Captain Palmer and Joseph Henry agreed on one fundamental point: Benjamin Gould was a significant part of the problem in Albany.

As September drew to a close, J. V. L. Pruyn commented in his diary that the Dudley Observatory controversy continued with great warmth. John Pruyn was now staunchly behind the Scientific Council; references in the trustees' pamphlet to his earlier support for Dr. Peters upset him. He conferred briefly with James Hall before launching himself into the local newspapers to establish his position in the quarrel. Hall implored Pruyn to say nothing that would draw the geologist into the conflict. James Hall had actively supported Peters's disastrous appointment also, but was now wary of any move that would pull him further into the acrimonious dispute. The best men of the city, he said, were splitting into two warring camps, "arrayed against each other with the most bitter and unrelenting animosity and all for science, as they say." Hall feared it would take years to re-establish harmony and that the "animosities engendered here will be elsewhere manifested."[16]

As if to add solely to the heat in the last days of September, another scorching pamphlet came roaring off the presses to attack the trustees. The author, who signed himself as "Observer," was a staunch Democrat and close business associate of Erastus Corning. "Observer" was the hot-tempered George Hornell Thacher, a supplier to the New York Central and vice-president of Corning's bank. Thacher had his own vendetta with Olcott. He accused the bank president, who was also president of the prestigious Albany Rural Cemetery Association, of making "cruel remarks" about the death of Thacher's daughter. Thacher launched a frontal assault on Olcott's reputation. The bank president was accused of gross incompetency and continuing recklessness in managing the observatory's funds. No more damaging attack could be made on an antebellum businessman, especially a banker who was entrusted with the fortunes of others. Reputation was everything. In contrast with this denunciation of Olcott, Thacher praised Erastus Corning effusively for his honest character and wisdom in money matters. Thacher continued a line of attack begun by the Scientific Council in their *Defence of Dr. Gould.* Why did the trustees refuse to make a public accounting of the observatory's financial condition? As businessmen, Thacher charged, they looked ridiculous; their reputations were already ruined. With such mismanagement how could the Dudley Observatory attract future donations? "Observer" attributed all the actions of the trustees to Olcott and accused him of crushing anyone who opposed him. Thacher later wrote that he used a pseudonym because he feared the "*unrelenting vindictiveness*" with which Olcott pursued anyone who crossed him. Thacher's pamphlet circulated widely but, because of its polemics, no one was terribly impressed. He continued to insert articles in the local newspapers through September and October, some under his own name and others under the pseudonym of "Albany," aiming his blows consistently at Olcott's mismanagement of the observatory funds with which he was entrusted. Thacher's vehemence in attacking Olcott offers ample evidence that old rivalries within the Albany business community—centered around Corning's and Olcott's banks—played a less apparent but certainly important part in the divisions within the city over this conflict.[17]

If Olcott was taking his lumps from Thacher, Bache had some in store from the solicitor of the treasury. On 1 October Hillyer sent a report to Secretary Cobb. Hillyer explained that he had asked Bache in August, under the secretary's instructions, to provide specific information about the rights and interests of the federal government in the Albany controversy. He enclosed a copy of Bache's tardy reply, saying that he was not convinced that the government had rights that could be sustained in court. Bache's attempt to circumvent Hillyer's uncomfortably explicit questions about the terms of the agreement had not gone unnoticed. Hillyer was not impressed. Nor had the district attorney provided any evidence to support the Coast Survey's claim to possession of the observatory. The solicitor then pointed out that although Bache's course of action might involve lengthy legal entanglements, Cobb did not need a legal justification to comply with Olcott's request. Olcott had not, after all, invoked the law, but simply asked the secretary to use his discretionary power to order the Coast Survey out of the observatory. Hillyer suggested that the simplest way out would be to hand Bache his marching orders rather than try to sustain such a nebulous and politically dangerous claim in court.[18]

At this juncture the secretary of the treasury grew tired of the niceties of intradepartmental fencing and penned a blunt message of his own to Bache. He forwarded Hillyer's report and insisted that Bache supply firm answers to the questions he had avoided—at once. Howell Cobb stated his position succinctly. On the one hand, he was told that the Coast Survey was depriving a private corporation of its rights; on the other, he was told that the government had legal rights the trustees were trying to defeat. As secretary of the treasury, he refused to defer to either Bache's or the district attorney's legal opinions. Cobb insisted that all the facts be laid before him, so that he could make a decision himself, one "for which I am willing to be responsible."[19]

As the secretary's forceful letter made its way north to Bache, the district attorney sent a lengthy opinion to the treasury, enclosing a list of his expenses. The Treasury Department was now faced with paying out hard cash from a tight budget to support Bache's claim to the Dudley Observatory. The district attorney had concluded that the Coast Survey was a tenant for a term at the observatory but confessed that much of his information had come from Bache and Gould. He requested further authorization for action from the secretary. Further authorization meant that additional legal costs would follow. It was well known in Washington that Howell Cobb was a tight-fisted secretary of the treasury, loathe to commit to unnecessary expenditures.[20]

Bache was understandably alarmed. He wrote first for advice to one of his most trusted friends in Washington, former secretary of war and current senator from Mississippi, Jefferson Davis. Bache was afraid Joseph Henry was about to withdraw from the fight. Bache asked Davis to meet with Henry to bolster his martial spirit. He also suggested that Davis might call upon the secretary of the treasury, a fellow southerner, to offer words of support for Bache's position in the observatory dispute.[21] Bache then sat down to compose a careful answer to Secretary Cobb.

The Bache papers at the Library of Congress contain a long rough draft of this letter, filled with delicate rephrasings, careful deletions, and deft circumlocutions. The Dudley Observatory was not referred to by name. The occupied premises were described to the secretary instead as a longitude station situated on Van Rensselaer Hill in Albany. Bache stressed that this was the first time that the Coast Survey's right to occupy a site had been contested, but retreated from the sweeping legal opinions he had offered to Solicitor Hillyer in September about the extent of federal authority in the matter. At this point he was careful to say that he did not know who had legal title to the property but suggested that control was vested in the director, who acted under authority from the trustees. Then, in a nimble piece of footwork, Bache argued that his agreement to use the site had been amicably made with the president of the trustees and with the director of the Observatory. This director was not named. Later in his letter, Bache referred separately to Gould, not as director of the Dudley Observatory, but only as his assistant in charge of longitude work in the survey. Without a list of the players in this Albany scrimmage, an uninformed reader of Bache's letter would believe that the director of the observatory and the Coast Survey assistant were two separate people. How well informed Cobb was at this time remains unclear.

Bache's statement to Cobb that legal dominion was probably vested in the trustees was, of course, at variance with the claims of the Scientific Council. In June the scientists had charged that the trustees did not have the authority to dismiss Gould without the approval of the Scientific Council, nor did they have the authority to evict the Coast Survey. Now, reporting to his own superior, Bache avoided such claims. No mention of a Scientific Council was made in any correspondence with the treasury officials. No mention was made to Cobb that the Scientific Council sat in judgment on the actions of the trustees. The most difficult departmental ground—Bache's relations with Solicitor Hillyer—was crossed quickly and carefully. Bache insisted that he intended no discourtesy to Hillyer and emphasized that he had relied on the advice of the district attorney and members of the Board of Regents in his actions. He ended with the hope that everything was now clear to the secretary.[22]

In mid-October Bache turned his attention to Palmer's descriptions of the persuasive effect of the *Statement of the Trustees* in Washington. He was chafing to get out some reply to counter the damage, but Henry and Peirce each refused to take the time to prepare a response. At Bache's insistence Henry did write a brief draft in answer to the trustees' pamphlet, but Bache thought it too weak. Privately, Bache resented what he saw as Henry's loss of heart and vowed his own determination to see the battle through to the end. But Henry's increasing withdrawal from the conflict would sorely diminish the scientists' influence in both Washington and Albany. Gould was in no condition to be of any substantive assistance. The astronomer had returned to Albany in a condition Bache described as "physically incapable."[23]

Later that week Jefferson Davis offered his own tactical suggestions. At first Davis had been of the opinion that the scientists should offer a point-by-point refutation of the trustees' pamphlet; upon reflection he concluded that few in Washington would bother to read a lengthy document. The *Statement of the Trustees* should be left unanswered. Meeting the trustees in argument elevated the Albany men in the public mind and decreased the stature of the scientists. Peirce concurred with Davis, cautioning Bache that by continuing the pamphlet war they ran the risk of losing important friends: "Are we prepared for this? Would anything compensate it? I think not." Both friends feared that the Albany quarrel might lead to serious injury of Bache's position at the Coast Survey. Peirce also insisted that they had no right to drag Joseph Henry "as a sacrifice to Gould." Was this an indication that Peirce, like Joseph Henry, was thinking of pulling back? After all, as Jefferson Davis put it, the Scientific Council had "done all which friendship and chivalry required for Dr. Gould." Was it wise to risk high stakes on the table in Washington on a questionable hand in Albany?[24]

The Washington end of the game was clearly critical. On 14 October 1858, Captain Palmer of Bache's Washington office had a long meeting with Secretary Cobb. The secretary discussed the difficulty over the observatory at some length, Palmer reported. Cobb's attention had been intensified by a second letter from Thomas Olcott. Olcott stressed the trustees' desire to cooperate with the Treasury Department or any other governmental agency wishing to use the Albany facilities. But, in an intricately phrased sentence, Olcott indicated that the trustees now looked directly to the secretary of the treasury to return the institution to them. The Dudley Observatory was "now violently withheld by Professor Bache[,] Chief of the Coast Survey[,] and his subordinates."[25]

By stressing the trustees' willingness to cooperate with the government once Bache was removed from the observatory, Olcott was attempting to exert additional pressure on Bache through his superiors. But the labored phrasing in Olcott's letter seems to have been misread by the secretary, who, as a southerner, had many other things on his mind in 1858. Cobb inferred that the trustees had no objections to Bache's completing the longitude observations. In contrast, my own reading of Olcott's letter indicates that the trustees wished to cooperate fully with any member of the government *except* Alexander Dallas Bache. On one issue, however, Cobb was completely certain. The Treasury Department would not be subjected to legal fees unless absolutely necessary. Cobb sent Bache a firm message. The treasury budget would not be tapped to pay the Scientific Council's legal expenses unless Bache could prove conclusively that the government had an unmistakable interest in the observatory and could support his claim with hard evidence.[26]

After a weekend's opportunity for rest and reflection, Joseph Henry opened his Monday morning mail to find another strongly worded message from Bache insisting that the members of the Scientific Council meet at once to draft a public reply to the trustees' pamphlet. Henry refused. His own problems were weighing too heavily. To complete his distress, Henry had been severely ill since his June visit to Albany. Joseph Henry was a quiet man, not inclined to exaggeration, but he described his illness as "a nervous affection of the head" that prevented him, in his words, from "using my mind as I could wish and indeed I have felt constantly that I was on the verge of breaking down." Henry refused to leave Washington to engage in another fight in Albany.

Joseph Henry agreed with Jefferson Davis. The strong impression made by the *Statement of the Trustees* would have faded by the time the scientists were able to publish an answering pamphlet. By printing a rebuttal, the scientists would add to the public impression that they had rushed into the quarrel as partisans of Gould rather than as men of principle. A heated reply would bolster the popular opinion that the scientists were "usurpers of the Observatory," neglecting their work for the government in order to "control from the love of power and personal aggrandizement, the science of the country." Henry was equally adamant that Gould could not participate in any communication the Scientific Council made in the future. He was explicit: "I object to his joining with us in any statement which may be made." Joseph Henry did not think that Benjamin Gould could remain above "personalities" and had earlier observed that Gould's inclusion in the Scientific Council weakened their position. How could Gould sit in judgment on his own actions and, at the same time, preserve any semblance of impartiality? Henry insisted that he would not allow his friendship with Gould to influence his judgment. He was particularly incensed by Bache's charge that his withdrawal was caused by new attacks on his leadership at the Smithsonian. Henry proudly affirmed that he intended to defend his record as secretary of the Smithsonian on its own merits. The Albany quarrel played no part in that record, and he resented Bache's remarks that "the enemy" had successfully separated him from the ranks. He insisted that no outside influence had affected his course, only his own belief in what he thought was right. Henry agreed to meet Bache in early November to draft a rebuttal but refused to jeopardize interests of greater importance than the Albany observatory just to please Bache.[27]

Jefferson Davis confirmed Joseph Henry's adamant refusal to involve himself further in the Albany struggle. Henry had spoken freely to the senator, insisting that he would "give no more to friendship than to justice." He would only do what he thought was right. As Jefferson Davis communicated these sentiments to Bache, he added the disheartening news that Benjamin Peirce was of a similar opinion. Although he supported Gould, Peirce had serious misgivings. He had earlier written Bache that Gould had been at the bottom of all the difficulties in his life and he did not wish to permit him to disturb the remainder of his days. Both Henry and Peirce each told Davis privately that they did not feel it "incumbent upon them to proceed further for the defence of Dr. Gould."[28]

Jefferson Davis was a loyal friend. He visited with Erastus Corning to discuss Bache's concerns and then went to see Howell Cobb on the following Wednesday, 27 October. The secretary of the treasury assured the senator that the Treasury Department would sustain all of Bache's legal rights at the observatory. This encouraging message was dampened by Cobb's subsequent statement that Solicitor Hillyer was currently preparing an opinion on the legality of Bache's claim. This was not welcome news. Bache had justifiable qualms about Hillyer's willingness to commit the resources of the Treasury Department unless forced to do so. The legal grounds Hillyer demanded were simply not available. Bache had evaded Hillyer's probing questions about the exact nature of the contract by which he held possession of the observatory because he knew the informal agreement would not convince the doubtful solicitor. Hillyer was not likely to commit the treasury to defend Dr. Gould nor to support the Scientific Council. Before he started home to Mississippi, Senator Davis visited with a member of Bache's staff to say that although Howell Cobb seemed favorable to Bache's position, he was a careful man. Bache's fears were sound.[29]

Hillyer examined Bache's report meticulously. He reduced the complexities to two legal questions: Did the superintendent of the Coast Survey have a binding contract with the trustees? In the absence of such a contract, did the laws of Congress authorize forcible occupancy of an observatory in Albany? The solicitor reviewed the letters describing the oral agreement and concluded that the January 1858 proposition had been mutually advantageous to both the Coast Survey and the trustees; but that agreement had never stipulated that the trustees agreed to give up their right to possess the observatory. Neither Bache nor the district attorney offered any evidence justifying a forcible occupation of the premises against the wishes of the trustees. Hillyer concluded that Bache could not base his claim on the terms of January's oral agreement. In reviewing the second question—whether the Coast Survey could forcibly possess property against the wishes of its owners—there were no precedents. In Hillyer's judgment only the shack the survey erected to house its instruments might be held until the observations were completed. The trustees had not objected to construction of the shanty. Because of this implied consent, the Coast Survey might be able to press its claim to maintain possession of the shack until the observations were finished. The solicitor was insistent that his opinion on the rights of the Coast Survey should neither be misunderstood nor expanded by Bache to cover other claims. As far as Hillyer was concerned, the Treasury Department would support Bache's right to sit in the tiny, temporary shack, about five-by-nine feet, that housed the longitude instruments used by the Coast Survey. The claims of the federal government ended there. The survey crews had no legal right to use the observatory building, any of its instruments, the dwelling house, or the grounds of the Dudley Observatory.[30]

The trustees must have chuckled over the news. The Washington end of the battle had indeed been critical. Bache had been defeated by his superiors in the Treasury Department. Political sensitivities to any arbitrary use of federal power were high in 1858. As one editor observed, the solicitor's opinion was proof that federal officials, no matter how exalted, could not invade an independent state to seize private property for public purposes.[31]

It is obvious that the conflict that erupted in Albany over the Dudley Observatory was not a simple story, but a Byzantine tangle of conflicting ambitions and centers of power—in the scientific community, in local business, in local politics, in state politics, within a federal department. Individual motives wound like intertwined threads through all levels of governmental and scientific institutions as well as through the intricate competitive patterns of the local business community. The conflict has been described as a battle between altruistic scientists and provincial trustees, but as the strands of individual intentions are pulled, they tug at vast networks of interconnected interests that defy such simplifications. Bache's actions had their sources in motives besides scientific selflessness, especially with regard to Dr. Peters and control of the observatory while the heliometer was still a possibility. Thacher's hatred of Olcott cannot be attributed to an idealistic vision of pure science nor to an empathy with Dr. Gould. His rage flowed from sources deep within local rivalries in Albany. Mrs. Dudley's attack on the trustees has to be viewed with an understanding of her nephew's interest in overturning her contribution to the endowment. Corning's opposition to Olcott involved many issues beyond science. Such substrata of interests, emotions, and pride—on both sides—cannot be ignored in seeking an explanation for the bitter depths of this conflict.

CHAPTER 12

The Last Campaign

Political parties in New York had splintered under the stresses of the summer conventions of 1858. This fracturing of traditional political alliances combined with the heated quarrel over the observatory to make the November election in Albany more acrimonious than any ever seen before. J. V. L. Pruyn confessed to his diary that he had worked vigorously in the campaign because of his anger over the observatory dispute. Pruyn observed that most of the trustees had worked against Erastus Corning, especially Olcott, who was "understood to have been particularly active and no doubt contributed largely in money to the expenses of the election."[1] Despite the strength of his opposition, Congressman Corning won handily. Olcott's close friend, Judge Amasa J. Parker, was also defeated in his second effort to be elected governor of New York. Pruyn noted that Olcott, formerly a Democrat, was now deep in the Republicans' secrets. Two years later Olcott would oppose Corning openly as both men vied for the congressional seat in the 1860 election. Corning, a Democrat, would win that election also.

Despite victory for his friends in the 1858 elections, Bache was increasingly frustrated. As superintendent of the Coast Survey, he was enduring scathing attacks in the press for his attempts to use federal power in the observatory quarrel. Rumors of Solicitor Hillyer's opinion prompted biting criticism of Bache's effort to have the taxpayers fund his legal expenses in the Albany dispute. His use of the franking privileges of the Coast Survey to mail out the Scientific Council's pamphlets along with those of George Thacher was attacked in the press. Bache complained bitterly to Peirce of the lack of support he was receiving from Secretary Cobb.

In his irritation over the trouble caused by matters in Albany, Bache refused a scholarly article offered for inclusion in the annual report of the Coast Survey by O. M. Mitchel. Bache insisted that Mitchel must give an open answer to the trustees' statements before anything would be accepted for publication. Joseph Henry proved exasperating too. Henry had been true to his word, meeting with Bache to draft a response to the *Statement of the Trustees*, but the two had been unable to agree on a reply both were willing to sign. Soon after this impassé, Henry received an upsetting letter from Peirce in which the mathematician accused him of deserting Gould under fire. Peirce had evidently overcome his own doubts and decided to continue the fight. Bache reassured Peirce that his letter to Henry was necessary, arguing that it was their duty to tell "Smithson" what his friends thought about his actions in the Albany conflict.[2]

Bache kept to a hard line, insisting that his own principles differed greatly from Henry's in deciding how to support friends, how to "carry on a war," and how to form personal opinions. When Henry and Bache met again, the "Chief" told the agitated secretary of the Smithsonian that he was not willing to sign any of the statements Henry had drafted. None of them were strong enough to do any good. Henry was upset, but Bache concluded that "Smithson's" mind remained fixed on Morse rather than upon the machinations of Olcott. During November Bache was immersed in the details of preparing his official Coast Survey report for its final printing, but Olcott was ever on his mind. The *Statement of the Trustees* was circulating widely, sent by the banker to everyone who requested a copy. Certainly, "the judicious Olcott" and Dr. Armsby must have been the subjects of many ribald toasts when the Lazzaroni gathered for dinner that November.[3]

Gould missed the dinner. He had wrapped himself in seclusion at "Fort Dudley." Seldom venturing out, Gould described himself to James Hall's wife as "absolutely wretched with a severe attack of my head and nerves." The astronomer wrote that he was unable to do much and wished desperately to go home to Cambridge to have "'a regular sick' of it, but even this dim luxury was denied." Declining Mrs. Hall's Thanksgiving invitation, Gould promised to present his "peaked and somewhat dismal countenance" as soon as he felt up to a visit.[4] Gould did manage to go home to Cambridge for Thanksgiving, visiting with Peirce while he was there. He insisted that a weak reply to the trustees' pamphlet would be worse than useless. Bache agreed, but dejectedly concluded that it was probably too late for a rebuttal to do much good anyway. For that delay Bache blamed Joseph Henry.

The "Chief" quizzed Peirce on his impressions of Gould's health and state of mind over the holiday. Bache had made his own earlier assessment of the distracted astronomer, concluding that Gould was "not well in body or in mind." Bache confided to Peirce that it was no surprise that Gould's troubles were overwhelming him, for "his health for some years has been wretched and as I think from mental causes." In the meantime Gould had returned from Cambridge to his wintry isolation in the observatory, concentrating on preparing his own reply to the trustees' pamphlet. He confessed to Peirce that he had become "desperate on the subject!"[5]

Bache was himself growing desperate. Joseph Henry's refusal to put his name to any statement from the Scientific Council that would further enflame the controversy had effectively quashed any response to the trustees' pamphlet. With this avenue closed, Bache turned to a casual suggestion made earlier in the autumn by Jefferson Davis. Davis had commented that he did not see how the Scientific Council could resolve the dispute except by arbitration or a lawsuit. The solicitor's opinion ruled out a lawsuit financed by the Treasury Department. Jefferson Davis had judged arbitration to be impracticable when he mentioned it to Bache in October. At this point Bache decided the idea was worth a try. Arbitration had already been suggested by the Scientific Council earlier in the summer. This time, Bache and Peirce arranged to have the request come to the trustees from some of the donors to the observatory. A petition asked the trustees to submit the altercation to five respected arbiters. The trustees rejected the proposal.[6]

With true Victorian doggedness, Bache refused to concede without a struggle. Could the negative decision presented by the solicitor of the treasury be circumvented? On 16 December 1858, Bache wrote to James Black, attorney general of the United States, to ask if the Coast Survey might remain in possession of the observatory property through the government's power of eminent domain. The attorney general's answer would not be received until February.

The festivities of the Christmas season swept all else aside, even the dispute over the observatory, for everyone but Benjamin Gould. Living like a recluse on the icy hill, he was afraid to leave the premises lest the trustees seize the property in his absence. He was also worried about his father, who was ill in Cambridge. Gould filled his hours by working intensively on his own lengthy pamphlet attacking the trustees. Gibbs wrote Bache to say: "Poor fellow . . . what a load of care and responsibility on top of his other troubles."[7]

Events were to be taken out of Gould's hands as the dispute moved to a dramatic climax. The observatory must have been in Olcott's thoughts over the holidays. Perhaps one of his New Year's resolutions was to bring this continuing source of irritation to an end. The wheels of bureaucracy were grinding too slowly in Washington. The date by which the scientists were supposed to vacate the "premises under the eviction notice had long past. Something had to be done. The bank president decided to terminate the dispute himself.

On 3 January, 1859 the first Monday morning of the new year, Olcott met with Elisha Mack, an Albany policeman. As president of the board of trustees, Olcott showed Mack an authorization from Ira Harris, justice of the State Supreme Court, which gave him permission to appoint Mack and "such persons as he may select to aid him, to take possession" of the observatory and all the buildings. Mack was "to remove therefrom all persons or property" that did not belong on the premises.[8] Although Gould later charged that fifteen to twenty-five members of a "gang of toughs" joined Mack, local reports indicate that Sergeant Mack chose seven men to accompany him. He arrived at the observatory gates at two-thirty that afternoon. In writing about the events later, Gould insisted that he had not been present when Mack and his troop arrived but was at a friend's house working on his pamphlet. In his own report, Mack stated that he was told by the young man who answered the door of the dwelling that Gould could not be disturbed because he was having lunch.

Mack and his men crossed to the observatory and entered the locked building through a window. They found two women washing clothes in the basement and reported that the meridian circle was standing in the open, covered in snow. The rooms of the observatory were described as being in a filthy condition, especially the basement, which appeared to have been used for raising chickens as well as for doing laundry. Mack added the intriguing statement that the basement had been used "for other purposes which I will omit to mention."

The doughty Mack left one man in charge in the observatory and returned to the dwelling house. He did not see Gould until four o'clock that afternoon, when he showed him Olcott's authorization. Gould asked if Mack was prepared "to repel force with force, if necessary?" Mack said that he was. Gould quickly replied, "Mr. Mack, we will have no 'fisticuffs.'" The astronomer asked for two days to pack, promising to make no resistance if Mack would leave the observatory premises. The policeman refused. Mack and his men remained in the house all night with Gould and his four assistants. The following morning Gould wrote Peirce that the very air he breathed was "a mixture of bad whiskey, worse tobacco, cheese, and Irishmen." Gould left that morning.[9]

When he attempted to return to the observatory the following day, the astronomer was met by Thomas Olcott, Judge Harris, and Dr. Armsby. They insisted that he leave at once. Benjamin Gould would not be allowed to enter the Dudley Observatory as director again. Dr. Armsby wrote to George Bond to ask if he would be interested in taking over the Dudley Observatory. Bond refused, saying that his father was ill, and he did not wish to leave the Harvard Observatory.[10]

In a letter to the *Albany Atlas and Argus* describing his eviction, Gould charged Mack and his deputies with being drunk and disorderly all night, "exercising the rule of brute force over the premises." Gould argued that he could have "met force with force" but instead chose to yield. Elisha Mack protested against this public attack on his reputation, saying that it was well known in Albany that he had not drunk intoxicating beverages for many years. Mack was a staunch supporter of the temperance movement. In contrast, the policeman charged that Gould knew very well "*who* threatened violence, and *who* had imbibed so much that it was visible to everyone present." According to Mack he helped Gould pack his belongings, and no harsh words had been exchanged. He was indignant at Dr. Gould's description of those who assisted him as a "lawless band of marauders" who had wrecked the house. Gould's description of Elisha Mack and his merry men has lingered in the literature of this conflict. It is an exciting picture to imagine—the astronomer thrown bodily out of his observatory into a snowbank and his belongings hurled after him—but it doesn't appear to match the facts of the story.[11]

The agitated astronomer rented a house in Albany and sat, surrounded by his assistants and his trunks. Describing himself as "pressed, and almost prostrated," Gould urged the principal donors to the observatory to petition the state legislature to revoke the charter because the institution was completely under the control of the iniquitous Olcott. He charged the trustees, especially Olcott, with a "depth of moral depravity which it is painful to contemplate."[12]

The tables had been turned. Bache had earlier argued that if occupation by the scientists was illegal, Olcott should resort to the courts. As the editor of the *Albany Atlas and Argus* now observed, the responsibility for entering the courts had been thrust upon the Scientific Council. Like all lovers of tales of action, the editor promised his readers that the eviction of Dr. Gould had not ended the controversy. More thrills were in store. He referred to the Scientific Council's effort to overthrow the original charter and noted that the trustees had already initiated a countermovement in Congress against the Coast Survey. As the editor described it, the conflict would now "take a wider range, and become national and involve political men in its entanglements."[13]

This prediction was correct, if somewhat late. Political influence had already been a weighty part of this struggle. Two days after Gould's unwilling departure from the observatory, Bache met with Secretary Cobb and wrote the attorney general of the United States again. Bache urged Attorney General Black to arrange a meeting with Congressman Corning to discuss the "forcible seizure" of a Coast Survey station by the trustees. The attorney general agreed to meet with Bache and Corning the following day. As he readied his forces, Bache turned once again to Joseph Henry, begging him to write something for immediate publication. Henry refused. He described himself as constantly harassed and told Bache that he could not possibly "give proper attention to the Dudley matter" until the following week. Perhaps Bache's staff could prepare a draft they might go over together, he suggested. Henry kept to the position he had previously taken in refusing to reply publicly to the trustees' pamphlet. As for the present flurry of activity in behalf of the Scientific Council, Henry thought it pointless. The next statement of the Scientific Council's position should be the last, he insisted, because he had never been able to see where the controversy would end. If the legislature overturned the charter of the observatory, Gould could not remain because his position would be too unpleasant. If the legislature upheld the trustees, "there is then an end of the matter." Henry's personal pride was not engaged in the Dudley Observatory. His detachment enabled him to see that Gould would be the focus of damaging hostility and criticism as long as he remained in Albany. What good for science could come out of an institution whose director evoked such animosities?[14]

Bache's pride, however, was still engaged. On receiving Henry's note, he wrote Peirce that he found communication with "Smithson" to be painful. Bitterly, he observed privately to Peirce that the troubles in Albany had enabled him to see who might be relied upon and who not. Relations between Joseph Henry and Bache suffered a breach in the strife over the Dudley Observatory and would never be the same again. Bache always resented Henry's refusal to exert his influence as freely as Bache expended his own political resources in Gould's behalf. In September of 1859 Bache expressed himself candidly to Peirce, saying that "Smithson came between himself and me when he showed the white feather, during the Dudley fight. No one *else* could have cast such a shadow as he did, gigantesque."[15]

In Bache's estimation Joseph Henry's influence and reputation in the two-front battle—Albany and Washington—was crucial. But if Henry chose to withdraw from the field, they would have to plunge ahead without him. For himself, Bache refused to concede or to accept an ignominous defeat. He reassured Peirce that he was holding his own in Congress and remained undaunted. In his bones Bache felt that his opponents were preparing another attack, but vowed to Peirce to meet it with unflinching strength, despite Henry's defection.[16]

Bache would continue his fight on "the Washington end." He had Gould write a report stating that Coast Survey instruments had been forcibly seized by the trustees. This was forwarded to the secretary of the treasury with a cover letter asking for official instructions from Cobb to guide Bache's actions. He must have hoped the secretary would feel compelled to respond to a direct provocation offered to government officers by the trustees. Bache also told the secretary that he was relying on advice from powerful "political men" in Washington. Apparently acting under such advice, he wrote Gould a formal letter in mid-January. In Gould's judgment, were the Coast Survey instruments liable to be injured? Bache sent copies of this correspondence to Senator Pearce of Maryland, asking if he would meet with Jefferson Davis on the matter. Although Davis was working hard during this month to have Stephen Douglas expelled from the Democratic party at the national convention, he and Pearce promised to confer.[17]

At this same time, Olcott, now in possession of the observatory, wrote again to the secretary of the treasury to explain the latest events. Though insisting that the trustees could no longer allow the property to be withheld, he assured Cobb that no one thought the secretary had sanctioned the "usurpations" by the Coast Survey. Olcott continued with a strong statement that placed the observatory quarrel firmly in the midst of some of the thorniest questions of 1859. He knew Cobb, formerly a powerful Georgia congressman, would be extraordinarily sensitive to questions of federal authority. Olcott insisted that only a case of "manifest right and urgent necessity" could justify "the invasion of a state by federal officers and the seizure of an institution chartered by its Legislature, ..."[18] After placing the conflict on such delicate state-federal grounds, Olcott then got to the main point of his letter. The transit instrument purchased by the Coast Survey three years earlier was still in its original packing crates in the Dudley Observatory. The telescope had been ordered by the Coast Survey for use at the Albany observatory with the understanding that the trustees would be able to purchase it when the survey had finished with it. Olcott reminded Cobb that the trustees had spent several thousand dollars purchasing piers for the instrument, provided a deposit to the makers, and paid all shipping charges. Surely the secretary of the treasury knew that the Coast Survey was expressly forbidden by Congress to set up such an instrument, which required permanent installation. The trustees wished to purchase the transit from the Treasury Department if a fair price could be agreed upon. Secretary Cobb passed Olcott's letter on to the superintendent of the survey for a decision. Not surprisingly, Bache refused. When he concluded his letter to Cobb, Olcott once again pledged the willingness of the trustees to cooperate with the Treasury Department, but suggested that the interests of the nation might be better served if the Coast Survey crew now occupying a house in Albany returned to the work for which they were responsible—surveying the coastal waters.[19]

If Olcott wished Gould out of Albany, the distressed astronomer matched the intensity of that desire with his own longing to be in Cambridge. His papers and personal belongings were in turmoil. In the midst of this upheaval, he was working frantically to finish his rebuttal to the trustees' pamphlet. Gould's fragile emotional stability had been swamped by the waves of troubles washing over him. He described his situation to Benjamin Peirce as "a terrible ordeal," saying that it was "only with great difficulty that I have kept to my work here."[20] Albany held nothing for him but reminders of the most upsetting experience of his life. He heartily wished to shake the snow of its streets off his feet for good.

During the first week of February, it looked as if Gould's luck might turn. William Cranch Bond, director of the Harvard Observatory, died at the end of January. Who would succeed him? The Harvard Observatory was the most prestigious astronomical post in the nation. Members of the Lazzaroni fellowship agreed that such a plum should not just fall into George Bond's waiting hands. Wolcott Gibbs wrote excitedly to Peirce that Gould alone should fill the vacancy, but cautioned that Gould must not be brought forward at all except to win, "in other words the thing must be certain before he goes into an election. . . . What a triumph for the Lazzaroni if he should succeed! Then those Albany scoundrels would gnash their teeth with impotent rage."[21] But Benjamin Peirce had long coveted the Harvard Observatory directorship himself. Peirce had earlier attempted to bring the observatory under his control by including it in the Lawrence Scientific School at Harvard. The Bonds had managed to avoid any inferior position during that time, and the observatory had been restored to its independent status within the corporation. Peirce confided privately to Bache that although Gould was "head and shoulders above any other man" in his scientific abilities, he was "wanting for the true qualities of an administrative officer." Bache agreed. Peirce's friends, including Louis Agassiz, began visiting members of the Harvard board of trustees to start Peirce's campaign rolling.[22]

Unaware of his mentor's ambition, Gould's hopes soared with the exciting possibility of becoming director at Harvard Observatory. The appointment would be a ringing vindication of his actions in Albany and the sufferings he had endured. When the news came to him from friends that Pierce was seeking the position, a dismayed Gould withdrew his own name immediately. Despite this premature parceling of prizes, George Bond was soon appointed to succeed his father as director of the Harvard Observatory. The timing of all this excitement could not have been more cruel for Gould. He was in no condition to bounce easily between such extremes of hope and despair. He took to his bed, writing Peirce that he was so ill that he was unable to do anything at all.[23]

As Gould sagged in despondency, the trustees of the Dudley Observatory experienced their own frustrations. To demonstrate the seriousness of their commitment, they invited various "scientific men" to Albany to discuss the future course of the observatory. George Bond asked to be excused, but gave his own assessment of the previous arrangements, noting that an "astronomer fit to direct its scientific operations" should not need a Scientific Council "and may find their control offensive or otherwise troublesome." Bond's dislike for the entire idea of a superior scientific body over the director is clear. In a letter to Franz Brünnow, Bond urged the trustees of the Ann Arbor Observatory not to commit themselves to a similar arrangement with a Scientific Council. He noted that something along the same lines was hinted at for the Harvard Observatory, but came to nothing. Sweeping assertions of professional authority were not part of George Bond's style. Instead, both he and his father before him maintained amicable relations with the official committee appointed by the overseers to visit the observatory. George Bond had their full trust and support because of his reasonable attitude and obvious competency.[24]

Although George Bond would not come to Albany to discuss the Dudley Observatory, he expressed his full confidence in O. M. Mitchel's abilities to bring honor to the institution. Mitchel had agreed to confer with the trustees about the observatory's future. A series of public lectures by the astronomer was planned that the trustees hoped would generate enough local enthusiasm to convince Mitchel to become director of the observatory. The trustees' intentions were serious and solemn in planning this convocation of "scientific men," but the results were farcical. Once again, the scenes of the Dudley drama seem to have been lifted from those of an operetta.

Mitchell arrived in Albany on the last day of January, intending to stay for a week. On the next morning, Olcott, Dr. Armsby, and several other trustees ceremoniously paraded to the hotel to meet with the astronomer. In the meantime the wily George Thacher was busily working to scuttle the carefully planned arrangements. Thacher visited with Gould and gathered excerpts from Armsby's letters in which the physician had disparaged Mitchel's abilities and work. Thacher then composed a letter to Mitchel, including many of Armsby's most damaging comments, offering to present the complete texts if Mitchel wished to read them. Thacher wrote Mitchel that this information would enable the astronomer to judge more wisely the men now asking him to take over the Dudley Observatory.

Mitchel met with the trustees at ten o'clock. By one in the afternoon, he had received Thacher's letter. He checked out of the hotel immediately and left on the next train to New York City. Within a few minutes of his precipitous departure, Dr. Armsby asked at the desk for Mitchel and was told the astronomer was no longer registered. Shocked, the indefatigable doctor ordered a sleigh and set out in hot pursuit of the train. As Thacher chortled to Corning: "There the curtain dropped. Whether Armsby is still chasing the Hudson River train, hallowing after Mitchel you can determine as well as I."[25] Dr. Armsby failed to catch the fleeing astronomer. The local newspapers pondered the mystery of Mitchel's sudden disappearance without giving any of his scheduled lectures. Olcott was furious with Armsby and made arrangements to meet privately with Mitchel in New York to smooth his ruffled feathers. George Thacher gleefully described the melodrama to Congressman Corning. Corning had a good chuckle, then forwarded Thacher's letters to Bache for his enjoyment.[26]

Bache was in need of a good laugh. Although the vaudevillian scenes in Albany might have eased some of his moments, he was experiencing serious difficulties in Washington. The Light House Board with which he was associated had severe funding problems in Congress. On 5 February 1859, Secretary Cobb sent him the district attorney's bill for legal services in Albany and said that it must be paid directly from Bache's own closely guarded Coast Survey budget.[27]

During the same week, Thacher was industriously scurrying about Albany. He met with J. V. L. Pruyn and Gilbert C. Davidson. Davidson was a business partner of Corning's and a trustee who supported the Scientific Council against the majority. Together, they drafted an application to the state legislature to overturn the original charter of the Dudley Observatory. Thacher warned Corning that if they were serious in this attempt, they had to be "systematic." Thacher believed that Olcott would force the powerful Thurlow Weed to sustain the trustees and would "pour out money like water." The combination of "Thurlow Weed and the almighty dollar" would effectively block any action against the trustees in the legislature. If the members of the Scientific Council intended to make a strong effort against the trustees, Thacher insisted that they had to send money to Albany. If the political game was to be played, the Scientific Council would have to show they were serious by placing their ante on the legislature's table.[28]

Corning sent Thacher's letter on to Bache at once and asked him for an answer. How much were the scientists willing to pay in cash to overturn the Dudley Observatory's charter? Bache replied evasively, saying he did not feel competent to advise Corning about how to handle a state legislature, but he did want to make a firm show of strength. Bache did not say where the money for this show of strength was to come from. Presumably, it would be supplied by the man who had been so successful in "influencing" the New York legislature before—Congressman Erastus Corning, the "Railroad King."[29]

Yet Bache's Washington defences were clearly crumbling. On Tuesday, 8 February 1859, the attorney general of the United States delivered his opinion on the Coast Survey's eviction from the observatory. Black had examined all the information provided by Bache and concluded that the superintendent of the Coast Survey had not been in possession of the property in Albany through any power he exercised as a federal official. As far as the attorney general was concerned, the dispute was a private one between the members of the Scientific Council and the trustees. Bache's activity as a member of the Scientific Council was in no way connected with his authority as superintendent of the survey. Black stated flatly that the secretary of the treasury had no obligation or right to use federal funds to protect Bache in his private contracts. The scientists' insistence the previous February that there was no official connection between the observatory and the Coast Survey and that no reference be made to Bache's title as superintendent of the Coast Survey was now turned against them.

Howell Cobb received the attorney general's opinion, which simply confirmed that of his own solicitor. The secretary sent the papers on to Bache, suggesting that they meet as soon as possible. On reading Black's opinion, Bache wrote immediately to Corning. Could the congressman visit the attorney general at once to see if something might be salvaged for the Scientific Council? Corning replied that he didn't know if "this ass has given the opinion as indicated," but would see what he could do. But the Washington battle was over.[30]

Warfare was still waging furiously, however, in Albany. The State Senate had suddenly decided to investigate Erastus Corning, Jr. Corning's son often handled "political problems" for the New York Central Railroad for his father, easing the way by lubricating the legislature. This thrust through the younger

Corning was obviously aimed at the elder. Both Corning's son and George Thacher attributed this new move in the legislature to Thomas Olcott. Thacher reported that the questions were difficult, but young Corning had handled himself well before the investigating committee. No harm had come from it.[31]

In the meantime Olcott and Armsby were reportedly occupied in circulating a petition asking Mitchel to return to Albany to give his lectures. Thacher wrote that Dr. Armsby was putting it about town that he had received a letter from Mitchel "of a forgiving nature," and that everyone was enjoying a good laugh at the doctor's expense. Thacher promised that there was more fun ahead as Gould's pamphlet was expected from the printer at any moment. Many who had seen it were overwhelmed by its size. Gould's *Reply to the Statement of the Trustees* was rumored to be over seven hundred pages. William Mitchell, father of Maria Mitchell and one of the Harvard Observatory's Board of Visitors, asked George Bond: "What can be a man's condition that is compelled to write 700 pages in his defence?"[32] Over eight thousand copies of Gould's *Reply* finally appeared in the third week of February; he believed himself completely vindicated.

March opened with the announcement that O. M. Mitchel would be returning to Albany after all. He gave his first lecture to a capacity audience in the First Congregational Church, despite a heavy storm. The following evening Thomas Olcott gave a large party in honor of the astronomer in his Arbor Hill mansion. At the close of his lecture series, Mitchel announced that he had agreed to come to Albany as director of the Dudley Observatory. His final hesitation was overcome when Mrs. Dudley made a remark during a visit that Mitchel and the trustees chose to interpret as an invitation to the astronomer to assume that position. The announcement infuriated Gould with its implication that Mitchel now had Mrs. Dudley's blessing.

Mitchel's appointment and Mrs. Dudley's seeming approval had a disastrous effect upon Gould. The following week he was again prostrate in despair. He described himself to fellow Lazzaroni, John Fries Frazer, as very ill and depressed. The distraught astronomer wrote a long report to Bache, urging a forcible seizure of the Dudley Observatory, either officially through the government's legal powers or unofficially by employing "private agencies." Gould assured Bache that if the authority were given to him, he would be happy to regain control of the observatory through the same forcible means by which Olcott had taken it.[33]

By this time Bache knew government action was impossible. Nor was he likely to hire a gang to wrest the observatory from the trustees. The criticism he had already undergone would be as nothing to that which would follow such an outrageous action. Besides, Bache had other problems in the spring of 1859. The Nautical Almanac's appropriation had been seriously attacked in Congress in March and April. His own budget had endured an "agony in the Senate" but emerged with its regular appropriation despite attacks from New York.

The Albany affair was over as far as Bache was concerned. One observer noted that "the clique" was discomfited by their defeat at Albany as well as by the rebuke Bache had been given by Secretary Cobb. Joseph Henry added his own final wish. In writing to James Hall later in the spring, Henry concluded his letter by saying that he hoped the Albany conflict had ended "and that my name will never again be associated in any way with that of the Dudley Observatory."[34]

CHAPTER 13

The Last Word

Joseph Henry may have washed his hands of the whole affair, but the trustees could not. They were still responsible for bringing the observatory into working condition. O. M. Mitchel had agreed to fill Gould's shoes as director of the Dudley Observatory but made it clear that he had no wish to follow that astronomer's footsteps into conflict. Once reassured by the trustees that they had no wish to repeat the experiences of the past several years, he accepted the position. Mitchel then wrote a conciliatory letter to Bache, explaining that he had been persuaded to accept the Albany position "at the earnest entreaty of many friends in New York City and in all parts of that state and elsewhere." He expressed his own deep regret for the divisions caused by the quarrel but insisted that he had not taken any part in the controversy. Mitchel also feared that, given the animosities aroused, the end of his own long struggle to secure a position would receive little sympathy from old friends—namely, Peirce and Bache.[1]

In addition to taking up the post in Albany, Mitchel retained the title of director of the Cincinnati Observatory, which had finally received an endowment to sustain it. Two transit instruments "of superior quality" had been ordered, one for each observatory. With the meridian circle in Albany, the great refractor in Cincinnati, and the two transits, Mitchel rejoiced that he would now have "four magnificent instruments . . . and the means to sustain as many capable observers—all to be devoted to the great work for which I have been ten years preparing and for which I am now ready."[2] Like Gould, Mitchel hoped to compile a catalogue of all the stars in the northern heavens to the tenth magnitude.

In Mitchel's absence Franz Brünnow—an immigrant astronomer with a doctorate from the University of Berlin and four years' experience as assistant director of the Berlin Observatory—was to be in charge at Albany. When Brünnow's name was mentioned for the Dudley Observatory, Gould was furious. Some of the letters whose publication had so embarrassed him during the conflict had been supplied by Brünnow. Because Brünnow had been one of the few astronomers to support the trustees against Gould and the Scientific Council, Gould held firmly to his opinion that Brünnow's actions were motivated by his desire for a post in Albany. Brünnow was clearly sensitive to the charges. When George Bond was appointed director of the Harvard Observatory, Brünnow took the opportunity to explain that the post in Albany was offered unexpectedly by Mitchel. Brünnow wrote that he had hesitated to accept the position for fear of giving the appearance of truth to Gould's allegations. "However I think that Gould's character is too well known on both sides of the Atlantic. . . . I hope it will save the Observatory from the mess to which the selfish course of the Scientific Council brought it. . . . I feel sorry for Gould, but I really think, that he and his friends have acted in a very foolish way."[3]

Brünnow would later play a central role in introducing German astronomical methods into the United States while at the Ann Arbor Observatory. Despite this methodological affinity, Gould opposed Brünnow's appointment as director of the Michigan observatory. His reasons sprang from both personal and nationalistic feelings. Gould believed that Americans, rather than Europeans, should fill American observatory positions. In addition to their animosity over the Dudley conflict, Gould and Brünnow had another area of rivalry. Like Gould, Brünnow published a professional journal, the *Astronomical Notices*, which attempted to compete with Gould's *Astronomical Journal*. While Brünnow was in Albany, the trustees agreed to underwrite the expenses of publishing his *Astronomical Notices* so that they would appear regularly.[4]

By December of 1859 a thirteen-inch refractor by Henry Fitz and the meridian circle were in full use at the Dudley Observatory. Ironically, the much-discussed heliometer, the instrumental magnet that had attracted Bache, Peirce, and Joseph Henry to the Dudley Observatory in 1855, was never built. As an astronomical instrument, the heliometer remained important through the rest of the nineteenth century, especially in the determination of the sun's distance in the 1870s, but observations were so slow and tedious that use of the instrument gave way inexorably to photographic methods. The "time ball" connections that Gould had described at the inaugural ceremonies were set up with the Customs House in New York City and the capital building in Albany soon after his departure. Both were activated by a telegraphic signal from the observatory each day at noon.[5]

Mitchel's wife, Louise, was in poor health, so he deferred his move to Albany until early May of 1860. As director of the Dudley Observatory, Mitchel hoped to concentrate the resources of the observatory upon the zone work required to formulate his long-planned catalogue. Unfortunately, once he arrived in Albany, close association with Dr. Brünnow did not prove compatible. Brünnow resigned in July. George Washington Hough was hired and would succeed Mitchel as director. Hough had a master's degree from Harvard and had worked for one year as an assistant at the Cincinnati Observatory.[6]

The Civil War soon overwhelmed all other concerns. Mitchel—a graduate of West Point—was never shy in his own behalf. The astronomer told Ira Harris that he wished to return to service but could not consent to serving in a subordinate position "or to find my superior in rank my inferior in ability and achievement."[7] Through Judge Harris's influence, Mitchel received a commission as a brigadier-general of the Ohio volunteers. On 18 August 1861 he left Albany for the last time. Louise Mitchel died two days later. George Hough remained in charge of the Dudley Observatory; Mitchel's friend, Henry Twitchell, had charge of the Cincinnati Observatory. O. M. Mitchel's military career was short but filled with the same flair he brought to the lecture platform. Commanding seventeen thousand men, he captured Huntsville, Alabama, in April of 1862 and was the commander in charge of the secret but unsuccessful raid immortalized as "The Great Locomotive Chase." Mitchel's successes led to his appointment as major-general in command of the Department of the South. He died of yellow fever at Beaufort, South Carolina, in 1862.[8]

George Hough served as director of the Dudley Observatory from 1862 until 1874, when he left to join the Dearborn Observatory. In Albany Hough divided his work almost equally between astronomical and meteorological observations. In 1872 Dr. Armsby organized a subscription campaign to purchase the instruments required for spectrum analysis, celestial photography, and magnetic observations. The following year the observatory united with the law school, the medical college, and other professional schools in Schenectady to offer postgraduate courses under the name of Union University of New York. Lewis Boss succeeded Hough as director in 1876.[9]

Boss was a well-respected astronomer. Before coming to the Dudley Observatory, he devised a set of uniform stellar declinations for use in the boundary survey between Canada and the United States. During this work Boss was impressed with the importance of producing a general catalogue of positions and proper motions to provide systematized information on the previous one hundred and fifty years of observations. He continued this task at the Dudley Observatory with little financial backing. He was finally able to convince the trustees of the Carnegie Institution of Washington to provide funding for the reduction of the great mass of observations accumulated between 1755 and

1932 to a properly weighted system. The Olcott Meridian Circle was dismantled and shipped to San Luis, Argentina, to complete the overlapping zone. The result was a great *General Catalogue* of the positions and proper motions of 33,342 stars, published jointly by the Carnegie Institute of Washington and the Dudley Observatory in Albany in 1938. Lewis Boss died in 1912; the work was completed by his son, Benjamin Boss, who succeeded his father as director of the Dudley Observatory.[10]

By the last decade of the nineteenth century, Benjamin Apthorp Gould, Jr., had made his own reputation at the National Observatory of Argentina. Gould was impressed with Lewis Boss's courage in undertaking the catalogue. Gould overlooked his earlier difficulties with the Dudley Observatory and helped Boss secure a grant from the Bache Fund of the National Academy of Sciences to continue his research on the proper motion of the solar system. In a manner befitting his lifelong dedication to the advancement of science, Gould continued to aid the Albany observatory that had been the scene of such traumatic personal experiences for him. By 1890 the more frequent and heavier engines of the New York Central Railroad that rounded the bottom of the observatory hill were causing vibrations that impeded telescopic observations. Lewis Boss arranged an exchange of property with the city of Albany. Benjamin Gould convinced Miss Catherine Bruce of New York City to donate twenty thousand dollars to the Dudley Observatory. By the time the move had been accomplished, Miss Bruce had been persuaded to raise her contribution to the Albany observatory to thirty-five thousand dollars.[11]

The new observatory was located on six acres formerly known as the Alms House Farm. The building was dedicated in 1893, again with full ceremonies, in the presence of the assembled members of the National Academy of Sciences. On Gould's death in 1896 the Dudley Observatory received his valuable personal library of astronomical books, including original editions of Copernicus, Galileo, Kepler, and other equally rare volumes. Later, Lewis Boss would take over publication of Gould's *Astronomical Journal.* The original observatory building, scene of so many fiery confrontations, stood abandoned on its hill until it burned in May of 1904.[12]

As James Hall had predicted, the hot tempers of the controversy took time to cool. In July of 1859 the geologist noted that prospects for science seemed dismal in Albany as a result of the observatory hostilities. Some of the animosities were reflected in local political campaigns. In April of 1860 George Thacher was elected to his first of five terms as mayor of Albany. In November Thomas Olcott ran against Erastus Corning for Congress and was defeated. With Lincoln's victory Olcott was offered the position of first comptroller of the currency but declined. During the Civil War, Olcott's bank served profitably as the chief subscription agent for government notes through an agreement with the new secretary of the treasury, Salmon P. Chase. In the spring of 1861 Judge Ira Harris replaced William H. Seward as senator from New York.[13]

Seward had been a friend to Bache and Joseph Henry. His replacement with the hated Judge Harris, author of the *Statement of the Trustees*, must have come as an unpleasant surprise. When news of the appointment reached Gould, he was so upset that he had to stop work for the day. He poured out his rage in a letter to Bache, calling Ira Harris a scoundrel, forger, liar, hypocrite, and villain. As ever, Gould laid the blame for Harris's appointment on the machinations of the president of the Mechanics' and Farmers' Bank, saying that Harris's nomination "must have cost Olcott a great deal of money."[14]

Although Gould had a successful future ahead of him, he was at one of the lowest points in his life when he left Albany. To compound his difficulties, financial problems forced the astronomer to take charge of the family business. He spent until September of 1860 trying to restore his "mental condition." The unhappy astronomer even purchased a horse and trotted around the Cambridge countryside "in the hope that the penance of the inferior extremity of the mortal coil, may propitiate the influences which dominate over the opposite, or cerebral pole of the system." In the aftermath of the Dudley conflict, Gould remained uncertain about undertaking any work, although urged to do so by Bache. Benjamin Silliman offered him a project, but Gould declined, saying "I am not yet mentally strong. . . ."[15] In October of 1861 he married Mary Apthorp Quincy, a distant cousin and daughter of the Honorable Josiah Quincy, former congressman and president of Harvard University. Mary Quincy Gould was a devoted wife, who had an obvious stabilizing influence on Gould's emotional states.

As Bache prowled the halls of Congress protecting his survey appropriation, it was inevitable that he would meet the new senator from New York—Ira Harris. The unexpected encounter came as both attempted to seat visitors in the crowded galleries of the House of Representatives for the fourth of July speeches in 1861. Bache shook hands with the senator and expressed surprise that Harris was "on the wrong side"—a pun that combined the senator's presence on the House (not the Senate) side of the Capitol with his opposition to Bache in the Albany controversy. Harris laughed heartily at Bache's sally, but as he moved away down the corridor, Senator Harris had the last word. Yes, he admitted that occasionally he was on the wrong side but, he continued in Latin, "*Spolia victori pertineant!*" The sentence was a familiar one in antebellum Washington, translating to the well-known political war cry: "To the victor belong the spoils!"

Harris's riposte demonstrated a ready wit, as he fielded a quick and humorous reply with several levels of meaning. That sentence, so closely associated with the "spoils system," had come into American politics with William Marcy, a member of the "Albany Regency" with Thomas Olcott. Marcy had been an influential national politician—secretary of war, secretary of state, presidential candidate—as well as a powerful governor of New York. William Marcy had also been a director of Olcott's Mechanics' and Farmers' Bank. The literal meaning of Harris's short Latin sentence reminded Bache that although he considered Harris to be on "the wrong side" in the conflict, the observatory remained firmly in the hands of the victors, the trustees. The connotations wrapped around the short sentence also reminded one who was knowledgeable in politics that some members of the board of trustees in Albany—like Thomas Olcott and Ira Harris—were old hands at wielding political influence in Washington.

Bache laughed to himself at the senator's adroit response. The frustrating struggle over the Dudley Observatory was indeed finished. It was time to move on to other concerns. Admitting defeat, Bache related the story to Peirce, with his own wry conclusion,

> My neck I bowed beneath the yoke
> Slain by the Judge's off hand joke.[16]

CHAPTER 14

Reflections and Conclusions

Posterity endorsed the opinions of the scientists in the conflict over the Dudley Observatory. Their version was first preserved in the eulogistic memoirs and biographies of the scientists who had been involved in the quarrel. From these origins the story moved out into the literature charting the history of science in the United States. The trustees won the battle but truly lost the war; they lost the attention of the future. Their interpretation of events faded out of the narrative, for few were inclined to delve into the yellowing pages of the densely worded and intensely argued pamphlets generated during the conflict. Besides, the story made a wonderful anecdote, graphically illustrating the problems facing scientists and other professionals in the egalitarian ethos of mid-nineteenth-century America.[1]

Because scientific institutions and the authority of experts are so much a part of present-day life, late-twentieth-century readers are more familiar with the lines of the scientists' arguments than those of the trustees in the Dudley Observatory conflict. Men like Bache, Peirce, and Joseph Henry were the nineteenth-century founders of the highly organized and disciplined "community of inquiry" we associate with modern science in the United States. The goals articulated and the institutional forms they espoused during their careers literally transformed the American scientific community. Their efforts to coordinate and shape that community and its institutions have an outline whose structure is known and whose results are acknowledged. Their arguments sound modern.

The trustees, in contrast, seem quaint. Their ethic of stewardship, with its emphasis on personal responsibility and character, provides a sharp contrast to many of the assumptions of daily life today. The increasing impersonality of present-day interactions makes the assumptions by which the trustees operated seem far removed. However grudgingly, modern Americans have adjusted to lives in which most services are provided impersonally, in which anonymous clerks confer or withdraw credit ratings for computerized cards, in which the bank teller is a machine, and even charitable donations are systematized and depersonalized by the United Way. The highly personal manner in which men like Thomas Olcott managed their affairs has given way to the formalized patterns of a bureaucratized society.

In addition to this obvious contrast, the trustees' spur-of-the-moment enthusiasm for vast new undertakings—setting up universities or observatories—also tends to distance them emotionally from twentieth-century observers. Although universities and the kinds of scientific institutions they wanted to establish would be the staging areas for the professionals of the future, the contrast between the trustees' informally personal style and a twentieth-century perspective is substantial. This difference is clear in their attitude toward establishing an observatory.

Today, laymen view astronomy as a highly specialized field of research, one supported by governments and undertaken by trained professionals who publish in esoteric journals only they can read. Present-day scientific institutions, observatories included, are both complex and intimidating. Although astronomy has always attracted the attention of wealthy individuals as benefactors, it is probably difficult for twentieth-century readers to empathize with someone like Dr. Armsby, determined as he was to establish a world-class observatory with the latest state-of-the-art equipment in his home town. If today the local gentry of some community decided to build their own six-meter telescope by circulating petitions and gathering donations, the effort would seem ludicrous. In many respects, therefore, the trustees' actions are not as easy to understand as those of the scientists in the quarrel over the Dudley Observatory. Nevertheless, it is a mistake to assume that the trustees were opposed to professional values. The difference in perspective between the nineteenth-century trustees and the twentieth-century layman lies, of course, in assumptions about professionals and their roles within institutions.

An interesting comparison might be made between the Dudley Observatory and the early history of the Rockefeller Institute for Medical Research with regard to assumptions about institutional and professional roles. Both institutions included a scientific council and a lay board of trustees. One of the motivating factors behind the founding of each institution was an awareness that the United States lagged behind Europe in research. A few months after John D. Rockefeller agreed to an initially limited funding for the project, seven advisers were named who were as significant in reputation in their time as the members of the Scientific Council of the Dudley Observatory were in theirs. Like Bache, Peirce, and Henry in science, the advisers to the Rockefeller Institute have been described as men who modernized American medicine. But the difference in institutional relations was tremendous. Like the early days of the Dudley Observatory, funding was not lavish at the Rockefeller Institute; but the scientists worked constructively within the budgets available while making their needs known. The Rockefeller Institute advisers met frequently with Frederick T. Gates—described as chief philanthropic lieutenant and architect of the major Rockefeller medical philanthropies—to discuss financial arrangements and the progress of work. In contrast with the unhappy conflicts at the Dudley Observatory, scientists and trustees at the Rockefeller Institute maintained clear lines of communication between the lay board and the scientific staff. Simon Flexner, first director of the Rockefeller Institute, was described by Frederick Gates as a masterful executive who inspired devotion and loyalty in his associates and subordinates. Unfortunately, the same could not be said of Benjamin Gould at the Dudley Observatory. The critical role of the first director of any scientific institution is clear. Although this brief sketch cannot draw larger conclusions, it is clear that roles within scientific institutions had assumed substantive definition between 1858 and 1908, the years between the Albany conflict and the formal organization of the board of trustees of the Rockefeller Institute.[2]

In the intervening fifty years, deference to experts grew along with accelerating urbanization and industrialization. Professional authority increased exponentially as science provided a growing understanding of nature; overwhelming technical achievements legitimated the authority of the professionals to whom they were attributed. In the century between the 1850s and the late 1940s, so much deference was accorded to "the experts" that reliance on professionals has been described as a defining characteristic of industrial society, in fact, one of the most distinctive features of modern culture according to Talcott Parsons.[3]

By the middle of the twentieth century, scientists had become the archetypal experts in the public mind. As professionals, they dealt in mysterious information about the fundamental nature of matter—from atoms to DNA—and received in return unconditional support and funding on the grounds of national security and/or human progress. The difference between the Albany trustees and modern Americans was that twentieth-century trust in experts rested largely on the competency provided by graduate schools; that trust was further guaranteed and enforced by an infrastructure of peer review and licensing boards. Where antebellum Americans relied on personal knowledge and reputation, twentieth-century Americans relied on institutional certification.

The relatively small area of professional competency that Bache and Gould were staking out for scientists in the 1850s has become an enormous turf in the 1980s. In contrast, the personal standards of an ethic of stewardship on which antebellum businessmen like Thomas Olcott relied were swept away in the economic and social restructuring that followed the Civil War. Legal instruments, incorporation, and limits on liability were more dependable as sources of discipline in the new economic order than reliance on an antebellum ethic of personal integrity. Yet, as present-day laymen increasingly ponder a world shaped by faceless institutions, impersonal bureaucracies, and values described as "scientific," echoes of the trustees' arguments on stewardship and personal responsibility have reemerged in environmental groups, antinuclear movements, and public discussions about recombinant DNA research and technology. In the past twenty years, the rule of experts has been openly and aggressively challenged.

After a century of nearly unquestioning support for most things scientific, increasing public dismay with the arrogance of professional authority has resulted in an epidemic of malpractice suits, erupting first in medicine in the 1970s and spreading into other professions so rapidly that the entire insurance industry has descended into a chaos now described as a national crisis. Pronouncements of scientific boards like the Nuclear Regulatory Commission no longer enjoy an automatic acquiescence from the public. The budgets of formerly sacred institutions like the National Institutes of Health and the National Aeronautics and Space Administration have been caustically reevaluated and drastically trimmed. NASA—as dominant in the twentieth century as the Coast Survey was in the nineteenth—has come under scathing denunciation for the arrogance, secrecy, and distorted priorities of its administrators and its decision-making processes. These were the very criticisms leveled against Benjamin Gould by Thomas Olcott. Like the trustees' response to Gould in 1858, the earlier benevolent image of the professional is currently tempered by growing public restiveness under what is now described as the tyranny of the experts.[4]

In like manner the marriage between science and government, nurtured by Bache and phenomenally successful by the early 1960s, has spawned critics ranging from liberal to conservative in the political spectrum in the past twenty years. Although Bache and Joseph Henry saw the national and state governments as more secure sources of funding than voluntary efforts in the nineteenth century, many now ask different questions: if government underwrites science, can science choose its own direction? How extensively are scientific judgments interwoven and shaped by policy decisions or by military imperatives? Of course, in addition to extensive governmental support, science is heavily funded by private industry. Scientific research is big business. But those who have to cope with the by-products of processes developed by scientific professionals in corporate laboratories are frustrated by their inability to find anyone who will assess the extent of the damage, much less accept responsibility. Questions of professional accountability are more openly pressed and more avowedly political.

Two essential issues raised in the antebellum conflict over the Dudley Observatory were the questions about the accountability of the professional and the extent of his authority. Although these issues were explored on a very minor stage in antebellum Albany, they remain important today. Like the trustees, modern laymen are increasingly aware that reliance on professional authority is only valid when the expert is trusted as a person with integrity, as someone with the best interests of the institution and community at heart. The trustees lost their faith in Gould's integrity because he repudiated his responsibility for his decisions. Gould's secrecy also raised serious questions among the trustees about their own responsibility to the institution and to the public. The trustees lost faith in the members of the Scientific Council when these esteemed advisers appeared to act in their own interests rather than for the good of the observatory or the community. In the Albany conflict, only Joseph Henry saw the central problem faced by the scientists. As he observed, "Our might is in our right. Let it remain thus."[5] In antebellum institutions trust provided the only firm foundation for the scientists' authority. In the past twenty years, very serious questions about the extent of scientific authority and the degree of scientific accountability have been raised on a variety of issues—unnecessary surgery, the release of manmade organisms into the environment, or the morality of assessing the outcome of a "limited" nuclear war for the Pentagon, among others.

The past does not necessarily provide lessons for the present. In some ways, however, the disastrous early years of the Dudley Observatory do suggest some cautions. The image of scientific authority projected by Benjamin Gould in Albany in 1858—his arrogance, his secrecy, his insistence on unlimited funding, his refusal to answer "ignorant" questions—is not unfamiliar behavior in the twentieth century. Despite his seriousness of purpose, his intellectual abilities, his dedication to the patriotic aim of raising the level of American science, Benjamin Gould refused to communicate on equal terms with the people who supported the observatory.

Obviously, there were other scientists in the nineteenth century who navigated the perils of establishing new institutions more successfully than Gould. In pointed contrast with Gould's disastrous career in Albany, Bache was extraordinarily effective as superintendent of the Coast Survey precisely because he acknowledged and carefully tended his constituency, the Congress. Joseph Henry's career at the Smithsonian and the Bonds of the Harvard Observatory provide two other obvious examples. Even Ormsby Macknight Mitchel, whose career has been used as an illustration of the preprofessional style of science, convinced the Cincinnati Astronomical Society to close the observatory to visitors so that he could devote himself to serious astronomical work. In his discussion of the career of George Hale—the astronomer who established the Yerkes and Mount Wilson observatories—Freeman Dyson argues that one of the reasons Hale was so successful was that he recognized the importance of public support in creating and maintaining scientific institutions. Scientists do not operate in a vacuum. Institutions are cooperative undertakings, maintaining their vigor only when support is given willingly and in good faith.[6]

The trustees of the Dudley Observatory were Gould's clients, his constituency. In essence, Benjamin Gould failed at a fundamental task for professionals, one that is still vitally important in the twentieth century. Senator William Proxmire's "golden fleece" awards of the past twenty years have often been no more than unfair potshots at valid scientific projects, but they do indicate that scientists are frequently ineffective in making their case to either Congress or the public. Dismissing ignorant questions, as Gould did with the trustees, is no more effective in the twentieth century than it was in the nineteenth.

The tragedy of the Dudley Observatory conflict, melodramatic and farcical as it was at times, was that both elites had similar goals. Despite earlier interpretations of the quarrel as a conflict between elitist scientists and provincial trustees, it should be obvious that the trustees of the observatory were not opposed to the scientific aims or the professional values espoused by Bache and Gould. Both scientists and trustees hoped to create a major scientific institution at Albany. Between 1855 and early 1858 they had no deep-seated differences in their hopes for the observatory's future. The good intentions of both elites converged in the observatory in those years, but were torn apart on the twin questions of professional authority and accountability within the institution.

By attempting to expand his version of scientific authority to exclude the trustees from any viable role within the institution and by refusing to be held accountable for his actions, Gould destroyed their trust and the basis of his future support. It was not opposition to professional values but the particular elaboration of a professional role within a nineteenth-century institution that caused the conflict. Despite his own normally acute political abilities, Bache supported Gould's claims to an authority that went far beyond that enjoyed by other directors within comparable antebellum institutions, including himself at the Coast Survey. It could even be said that any director who emulated Gould's secretive and uncooperative behavior would have a hard time today.

Whether the reader accepts the suggestion that Gould's behavior may have been the result of a manic-depressive personality disorder or not does not really influence the larger conclusion that can be drawn from the early history of the observatory. Scientists are still courting disaster when they do not communicate with or "educate" the laymen involved in the funding of their institutions and research. This issue is even more important in the twentieth century when television and the press often make laymen feel they are "armchair experts" on all kinds of scientific issues. Scientific questions receive so much attention in the media that automatic acquiescence to professional authority by the public simply can no longer be assumed.

On more narrow grounds of scholarship, the early years of the Dudley Observatory suggest a cautious approach by historians to conceptual discussions of the process of professionalization in the United States. The literature on professionalism is vast, but scholarly discourse on the topic can be confusing and contradictory. Common definitions present the first obstacle. Professionalism is often defined in terms of a set of special skills or knowledge beyond the reach of laymen, and as the creation of an organized subculture (journals, associations, etc.) with its own system of values. In the clash of definitions, only the most general descriptions of professional attributes remain as neutral ground, often so stripped in meaning as to be inadequate in assessing specific situations. Analysis is also complicated by the fact that professional goals as well as what was meant by the term "professional" have changed over time. Transferring definitions from one historical context to another presents problems. Despite the bare bones of the terms on which agreement is possible, generalizations from a few examples have often implied an inevitability to the process and style of professionalization.[7]

Most of the early studies of the professions assumed that they were a positive force in society. Recent scholarship tends to follow George Bernard Shaw's view that professionalism is more likely to be a conspiracy against laymen, with emphasis on the striving for power, money, and status by an elitist few at the expense of the many. Although the Lazzaroni do sound conspiratorial at times, such a view is unconvincing when expressed in materialistic terms, especially when explaining the quasi-religious intensity with which men like Benjamin Gould, Bache, or

Joseph Henry dedicated their lives to a professional ideology. This line of agrument also fails to provide explanation for the extensive commitment made by laymen like the trustees to the achievement of professional goals, unless one is willing to follow a Marxist analysis emphasizing class solidarity, a model that hardly fits antebellum America.

Despite the modern-sounding rhetoric of professionalism that can be abstracted from the letters of antebellum scientists like Peirce and Bache, the conflict in Albany reflects a more idiosyncratic process, one affected by too many local and personal factors among both scientists and trustees to fit neatly into any simple model. If research pinpointed developmental change, as reflected in attitudes, institutions, and changing perceptions of roles among specific groups over time, then more understanding of the process of professionalization might follow.[8]

What is clear in the Albany conflict is the fact that the two elites who battled over the observatory based their judgments on differing assumptions that were not directly focused on professional aspirations: assumptions about gentlemanly behavior, about institutions, about responsibilities, and about reputations. Those assumptions, on both sides, provide a helpful guide to understanding how the professional ideology fit into the larger culture at a specific time. While accusing their opponents of dastardly behavior and justifying their own, both the trustees and the scientists displayed and highlighted their own particular expectations.

In assessing changing relationships between professionals and laymen in different contexts, philanthropy offers a promising area of inquiry. Through voluntary donations of both time and money, an individual has the opportunity to transform his own ideas into actual institutions. In its most basic sense, a large donation is an attempt to persuade an institution and the specialists who tend it to undertake a particular mission, one the patrons feel strongly enough about to support with money from their own pockets. Institutions that have their origins as voluntary undertakings provide opportunities for an assessment of cultural assumptions. Much of the scholarship on antebellum philanthropy has concentrated on higher education or efforts to help the less fortunate.[9] In voluntary undertakings with a more specific purpose, like the Dudley Observatory, the relationship of elites can be charted clearly in terms of particular goals. Differing perceptions of what is expected by either professionals or laymen can be addressed over a period of time. The conflict over the Dudley Observatory is intriguing precisely because both sides shared a common goal, demonstrating the degree of convergence of specific professional goals with those of the larger culture at mid-century. Unfortunately, the effort failed for many reasons, not least of which was the personality of Benjamin Gould. Rigorous professional standards in science, however, were not at issue. Because the quarrel was so public, it provides an opportunity for a more accurate historical assessment of the degree to which professional ideology was accepted and the grounds on which it was resisted at a precise time in antebellum America.

Whatever the final judgment on the early history of the Dudley Observatory, the questions raised in Albany on the eve of the Civil War are still pertinent, given our increasing dependence on experts and growing suspicions about their integrity and accountability in our own time. Can scientists be the sole judges of their own actions? Bache, Peirce, and Gould insisted that was so in Albany. Joseph Henry had serious reservations. The Dudley Observatory fell apart over this question. Although the disastrous early years of the Albany observatory seem far removed from the institutional complexities of twentieth-century science, the questions raised in antebellum Albany about the extent of scientific authority and professional accountability remain vitally important today. No matter how imposing the superstructure of modern professional authority, that vast organization of institutions and careers subsumed in any profession rests ultimately and finally on the public's trust. When that trust in the integrity of the expert is shaken or lost, the price in blighted aspirations and unfulfilled institutions is a high one.

Notes

The following names, manuscript collections, and publications have been abbreviated:

AC	Airy Collection, Royal Greenwich Observatory Archives
ADB	Alexander Dallas Bache
BAG	Benjamin Apthorp Gould, Jr.
BC	Alexander Dallas Bache Collection, Library of Congress
BP	Benjamin Peirce
DAB	*Dictionary of American Biography*
DSB	*Dictionary of Scientific Biography*
GC	Oliver Wolcott Gibbs Collection, Franklin Institute
HUA	Harvard University Archives
OC	Thomas W. Olcott Collection, Albany Institute of History and Art
OMM	Ormsby Macknight Mitchel
PC	Benjamin Peirce Collection, Houghton Library,
RC	William Jones Rhees Collection, Huntington Library
TWO	Thomas W. Olcott

Chapter One: Introduction

1. Simon Newcomb, *The Reminiscences of an Astronomer* (Boston: Houghton, Mifflin and Company, 1903), 80.

2. The term *elite* is used in a functional sense to refer to the decision-makers within the local context and does not refer to *class*. For the purposes of this discussion, elite encompasses the influential individuals in Albany, regardless of family position. As Ronald Story points out, in antebellum America the "upper class" concept is less helpful, in that members have social prestige and group solidarity but may not be the dominant influence in the community. Ronald Story, *The Forging of an Aristocracy: Harvard and the Boston Upper Class, 1800–1870* (Middletown, Conn.: Wesleyan University Press, 1980), xii, 6–7; E. Digby Baltzell, *Philadelphia Gentlemen: The Making of a National Upper Class* (Philadelphia: University of Pennsylvania Press, 1979), 5–7; Peter Dobkin Hall, *The Organization of American Culture, 1700–1900: Private Institutions, Elites, and the Origins of American Nationality* (New York: New York University Press, 1984), 2–4.

3. The board of trustees of the Dudley Observatory included the following men during the years relevant to the conflict:

Name	*Tenure*	*Occupation*	*Alignment*
James H. Armsby	1852–1875	physician	majority
Gilbert C. Davidson	1852–1875	merchant	minority
William H. DeWitt	1852–1873	merchant	majority
Ira Harris	1858–1875	lawyer/politics	majority
Alden March	1852–1869	physician	majority
Thomas W. Olcott	1852–1880	banker	majority
Ezra P. Prentice	1852–1861	banker	minority
Robert Hewson Pruyn	1852–1882	lawyer/politics	majority
Samuel H. Ransom	1852–1885	iron founder	majority
Joel Rathbone	1852–1861	stove founder	minority
John F. Rathbone	1858–1901	iron founder	majority
John B. Tibbits	1852–1857	minister/inventor	minority
Gen. Stephen Van Rensselaer	1852–1858	land owner	withdrew
Isaac W. Vosburgh	1852–1889	merchant	majority
Eliphalet Wickes	1852–1861	bank director	minority
John N. Wilder	1852–1858	merchant	majority

4. Thomas Bender, "Science and the Culture of American Communities: the Nineteenth Century," *History of Education Quarterly* 16 (Spring 1976): 63–77; Richard G. Olson, "The Gould Controversy at Dudley Observatory: Public and Professional Values in Conflict," *Annals of Science* 27 (September 1971): 265–276; Trudy E. Bell, "'Garblings and Perversions': The Dudley Observatory Controversy," *Griffith Observer* 38 (November 1974): 2–9; George H. Daniels, "The Process of Professionalization in American Science: The Emergent Period, 1820–1860," *Isis* 58 (Summer 1967): 160–162.

5. Robert H. Bremner, *American Philanthropy* (Chicago: University of Chicago Press, 1960), 43–44.

6. Edward Pessen, *Riches, Class, and Power Before the Civil War* (Lexington, Mass.: D. C. Heath and Company, 1973), 251–268, 277–278, 304; Edward Pessen, "The Social Configuration of the Antebellum City: An Historical and Theoretical Inquiry," *Journal of Urban History* 2 (May 1976): 267–306; Robert A. Dahl, *Who Governs? Democracy and Power in an American City* (New Haven: Yale University Press, 1961), 11–12, 25–30.

7. Paul Goodman, "Ethics and Enterprise: The Values of a Boston Elite, 1800–1860," *American Quarterly* 18 (Fall 1966): 438–439, 448; Clifford S. Griffin, *Their Brothers' Keepers: Moral Stewardship in the United States, 1800–1865* (New Brunswick, N.J.: Rutgers University Press, 1960), 52–55.

8. Donald Zochert, "Science and the Common Man in Ante-Bellum America," *Isis* 65 (December 1974): 449; Charles E. Rosenberg, "Science and Social Values in Nineteenth Century America: A Case Study in the Growth of Scientific Institutions," in *Science and Values: Patterns of Tradition and Change*, ed. Arnold Thackray and Everett Mendelsohn (New York: Humanities Press, 1974), 21; Merle Curti, *The Growth of American Thought* (New York: Harper and Brothers, 1951), 318–319.

9. Deborah Jean Warner, "Astronomy in Antebellum America," in *The Sciences in the American Context,* ed. Nathan Reingold (Washington, D.C.: Smithsonian Institution, 1979), 64; Bruce Sinclair, "Americans Abroad: Science and Cultural Nationalism in the Early Nineteenth Century," in *Sciences in the American Context,* ed. Reingold, 37–42; Nathan Reingold, "American Indifference to Basic Research: A Reappraisal," in *Nineteenth Century American Science: A Reappraisal,* ed. George Daniels (Evanston, Ill.: Northwestern University Press, 1972), 38–62.

10. Richard J. Storr, *The Beginning of the Future: A Historical Approach*

to Graduate Education in the Arts and Sciences (New York: McGraw-Hill Book Company, 1973), 31.

11. Burton J. Bledstein, *The Culture of Professionalism: The Middle Class and the Development of Higher Education in America* (New York: W. W. Norton and Company, 1976), 195, 323, 333; John J. Beer and W. David Lewis, "Aspects of the Professionalization of Science," *Daedalus* 92 (Fall 1963): 772–773; Hall, *Organization of American Culture,* 36–38.

12. Nathan Reingold, "Definitions and Speculations: The Professionalization of Science in America in the Nineteenth Century," in *The Pursuit of Knowledge in the Early American Republic,* ed. Alexandra Oleson and Sanborn C. Brown (Baltimore: Johns Hopkins University Press, 1977), 34; Eliot Friedson, "Are Professions Necessary?" in *The Authority of Experts: Studies in History and Theory,* ed. Thomas L. Haskell (Bloomington: Indiana University Press, 1984), 5; Laurence Veysey, "Who's a Professional? Who Cares?" *Reviews in American History* 3 (December 1975): 419–423.

13. John Lankford, "Amateurs versus Professionals: The Controversy over Telescope Size in Late Victorian Science," *Isis* 72 (March 1981): 11.

14. Roger Hahn, *The Anatomy of a Scientific Institution: The Paris Academy of Sciences, 1666–1803* (Berkeley: University of California Press, 1971), x.

Chapter Two: Albany

1. William Kennedy, *O Albany!* (New York: Viking Penguin, 1983), 3, 5.

2. New York State. Office of the Secretary of State, Hon. Joel T. Headley. *Census of the State of New York for 1855; Taken in Pursuance of Article Third of the Constitution of the State, and of Chapter Sixty Four of the Laws of 1855, Prepared From the Original Returns, Under the Direction of Hon. Joel T. Headley, Secretary of State, by Franklin B. Hough, Superintendent of the Census* (Albany: Charles Van Benthuysen, 1857).

3. *Albany Atlas and Argus,* 4 September 1856; 6 September 1856; Arthur James Weise, *The History of the City of Albany, New York, From the Discovery of the Great River in 1524, by Verrazzano, to the Present Time* (Albany: E. H. Bender, 1884), 480; Joel Munsell, *Collections on the History of Albany* (Albany, N.Y.: J. Munsell, 1865) 2:68–69.

4. Munsell, *Collections*, 2:340–343, 391–396; *Albany Atlas and Argus*, 1 November 1856; Joel Munsell, *The Annals of Albany* (Albany: J. Munsell, 1850) 1:361, 363; 2:69, 299; 8:313–315.

5. "Buckingham's Sojourn in Albany," in Munsell, *Annals*, 9:304–305; "Charles Mackay in Albany," in Munsell, *Annals*, 9:340–343.

6. Nathan Reingold, ed., *The Papers of Joseph Henry* (Washington, D.C.: Smithsonian Institution Press, 1972) 1:xix–xx, 15–52; James M. Hobbins, "Shaping a Provincial Learned Society: The Early History of the Albany Institute," in *Pursuit of Knowledge in the Early American Republic*, ed. Oleson and Brown, 117–150; Wallace Kenneth Schoenberg, "The Young Men's Association, 1833–1876: The History of a Social-Cultural Organization" (Ph.D. diss., New York University, 1962).

7. Frederic Cople Jaher, "Nineteenth Century Elites in Boston and New York," *Journal of Social History* 6 (Fall 1972): 66–69; Henry Christman, *Tin Horns and Calico* (New York: Henry Holt and Company, 1945), 13–14.

8. Michele Alexis L. Aldrich, "New York Natural History Survey, 1836–1845" (Ph.D. diss., University of Texas, 1974), 1–24; Wyndham D. Miles, "Public Lectures on Chemistry in the United States," *Ambix* 15 (October 1968): 142; Samuel Rezneck, "A Travelling School of Science on the Erie Canal in 1826," *New York History* 40 (July 1959): 255–269; Walter B. Hendrickson, "Nineteenth Century State Geological Surveys: Early Government Support of Science," *Isis* 52 (September 1961): 357; Samuel Rezneck, *Education for a Technological Society: A Sesquicentennial History of Rensseluer Polytechnic Institute* (Troy, New York: Rensselaer Polytechnic Institute, 1968), 6–8, 15; Christman, *Tin Horns and Calico*, 13.

9. "Albany Medical College," in Munsell, *Collections*, 2:219–229; George R. Howell and Jonathan Tenney, eds., *History of the County of Albany, New York, from 1609 to 1886* (New York: W. W. Munsell and Co., Publishers, 1886), 2:220–221; Cuyler Reynolds, ed., *Hudson-Mohawk Genealogical and Family Memoirs* (New York: Lewis Historical Publishing Company, 1911), 1:138–140, 197–198; 2:540–541.

10. James Hall, "The New York Geological Survey," *Popular Science Monthly* 22 (1883): 824–825.

11. John M. Clarke, *James Hall of Albany: Geologist and Paleontologist 1811–1898* (Albany: n.p., 1921), 178–179; Charles Lyell, *Travels in North America* (London: John Murray, 1845), 16–20; Albany, New York, Committee on the Celebration of the Two Hundred Fiftieth An-

niversary of the Granting of the Dongan Charter, *Albany: A Cradle of America* (Albany, N.Y.: Argus Company, 1936), 52.

12. Emma Rogers, wife and editor of William Barton Rogers, places the beginning of the Society of Geologists as early as September of 1819 at Yale. Herman L. Fairchild states that the first formal efforts toward establishing an organization took place in 1838 and 1839 at the home of Ebenezer Emmons in Albany. Emma Rogers, ed., *Life and Letters of William Barton Rogers* (Boston: Houghton, Mifflin and Company, 1896), 1:168–169; Herman L. Fairchild, "The History of the American Association for the Advancement of Science," *Science* 59 (25 April 1924): 365–366; Sally Gregory Kohlstedt, *The Formation of the American Scientific Community: The American Association for the Advancement of Science 1848–1860* (Urbana: University of Illinois Press, 1976), 61–99.

13. Cecil J. Schneer, "The Great Taconic Controversy," *Isis* 69 (June 1978): 173–181; George P. Merrill, *The First One Hundred Years of American Geology* (New York: Hafner Publishing Company, 1964), 594–614.

Chapter Three: A Scientific Elite Comes to Albany

1. Thomas Bender, "The Erosion of Public Culture," in *The Authority of Experts*, ed. Thomas L. Haskell, 88.

2. Kohlstedt, *Formation of the American Scientific Community*, 96–97; John D. Holmfeld, "From Amateurs to Professionals in American Science: The Controversy Over the Proceedings of an 1853 Meeting," *Proceedings of the American Philosophical Society* 114 (February 1970): 22–35.

3. I am indebted to Marc Rothenberg, acting editor of the Joseph Henry Papers, for his probing questions on the distinction between professional and nonprofessional as far as the Lazzaroni were concerned. Everett C. Hughes, "Professions," in *The Professions in America*, edited by Kenneth S. Lynn and the editors of *Daedalus* (Boston: Beacon Press, 1965), 5–6.

4. Robert A. Stebbins, "Avocational Science: The Amateur Routine in Archaeology and Astronomy," *International Journal of Comparative Sociology* 21 (March–June 1980): 47–48.

5. Faculty Records of the Lawrence Scientific School, 11 November 1862, 18 November 1862, 2 December 1862, HUA; A. Hunter Dupree, *Asa Gray* (New York: Atheneum, 1968), 315–316; Warner, "As-

tronomy in Antebellum America," in *Sciences in the American Context,* ed. Reingold, 67–68.

6. Reingold, ed., *Papers of Joseph Henry,* 1:xviii; Reingold, "Alexander Dallas Bache," *DSB* 1:363–365.

7. Edward Lurie, "Nineteenth Century American Science: Insights from Four Manuscripts, *Rockefeller Institute Review* 2 (February 1964): 18.

8. Margaret Rossiter, *The Emergence of Agricultural Science: Justus Liebig and the Americans, 1840–1880* (New Haven: Yale University Press, 1975), 50; Henry James, *Charles W. Eliot: President of Harvard University 1869–1909* (Boston: Houghton Mifflin Company, 1930), 1: 106.

9. Storr, *Beginning of the Future,* 29–30.

10. Alexander Dallas Bache, "Remarks," *Proceedings of the American Association for the Advancement of Science* 6 (1852): xli–lx.

11. Louis Agassiz to James Hall, 3 August 1851, as quoted in Clarke, *James Hall,* 192–93; Circular, William Cranch Bond Correspondence, HUA.

12. Robert Silverman and Mark Beach, "A National University for Upstate New York," *American Quarterly* 22 (Fall 1970): 701–713; Howard S. Miller, *Dollars for Research: Science and its Patrons in Nineteenth Century America* (Seattle: University of Washington Press, 1970), 74–87.

13. Frances Trollope, *Domestic Manners of the Americans* (New York: Alfred A. Knopf, 1949), 316.

14. Donald M. Scott, "The Profession That Vanished: Public Lecturing in Mid-Nineteenth-Century America," in *Professions and Professional Ideologies in America,* ed. Gerald L. Geison (Chapel Hill: University of North Carolina Press, 1983), 23–25.

15. Josiah Whitney to James Hall, 13 January 1851, 24 January 1851, 29 January 1851, James Hall Collection, New York State Library; Richard J. Storr, *The Beginnings of Graduate Education in America* (Chicago: University of Chicago Press, 1953), 60–67; Samuel Rezneck, "The Emergence of a Scientific Community in New York State a Century Ago," *New York History* 63 (July 1962): 228–229.

16. Benjamin Boss, *History of the Dudley Observatory, 1852–1956* (Albany: privately printed, n.d.), 9; "Benefactors of Education and Science," *American Journal of Education* 2 (1856): 593–594.

17. OMM to TWO, 11 September 1852, OC.

18. TWO to OMM, 19 July 1853, OC; Benjamin Apthorp Gould, Jr., *Reply to the "Statement of the Trustees" of the Dudley Observatory* (Albany: Charles Van Benthuysen, 1859), 37–38.

19. Dr. Armsby to BAG, 8 December 1854, as quoted in Gould, *Reply to the Statement of the Trustees,* 39.

20. Gould, *Reply to the Statement of the Trustees,* 40–41; Dudley Observatory, *The Dudley Observatory and the Scientific Council: Statement of the Trustees* (Albany: Charles Van Benthuysen, 1858), 7.

21. A. Hunter Dupree, "Central Scientific Organization in the United States Government," *Minerva* 4 (Summer 1963): 456–457; A. Hunter Dupree, *Science in the Federal Government* (Cambridge, Mass.: Harvard University Press, 1957), 102–103.

22. Benjamin Silliman, Jr., "Science in America," *New York Quarterly* 2 (1853): 443.

23. George Biddle Airy, "On the Method of Observing and Recording Transits, lately introduced in America," *Philosophical Magazine and Journal of Science,* 3d ser., 36 (1850): 142–150; *Monthly Notices of the Royal Astronomical Society* 10 (1850): 26–34.

24. George Bond to Asa Gray, n.d. [1859?], Memorandum on Samuel John Johnson, George Bond Correspondence, HUA.

25. John F. W. Herschel, *Outlines of Astronomy* (London: Longman, Brown, Green, and Longman, 1851), 519–553; Agnes M. Clerke, *A Popular History of Astronomy during the Nineteenth Century* (London: Adam and Charles Black, 1908), 10–51; F. W. Bessel, *Astronomische Nachrichten* 8 (1831): 397–426; 15 (1838): 365–66.

26. Reingold, "Reflections on 200 Years of Science," in *Sciences in the American Context,* 14; Bruce Sinclair, "Americans Abroad" in *Sciences in the American Context,* 40–42; J. M. Gilliss to BP, 22 October 1847, PC; J. M. Gilliss to Elias Loomis, 10 April 1848, as quoted in *Science in Nineteenth Century America: A Documentary History*, ed. Nathan Reingold, (New York: Hill and Wang, 1964), 135–136, 144.

27. William C. Bond, Memorandum, 1859, W. C. Bond Correspondence, HUA; BP to ADB, 9 August 1863; 13 September 1863, PC.

28. Dudley Observatory, *Statement of the Trustees*, 7.

29. Gould, *Reply to the Statement of the Trustees*, 44.

Chapter Four: Benjamin Apthorp Gould, Jr.

1. Rosenberg, "Science and Social Values," in *Science and Values,* 23.

2. The largest deposits of Gould's letters are in the Benjamin Peirce Collection, Houghton Library, Harvard University, the Bache

Collection, Library of Congress, and the Oliver Wolcott Gibbs Collection, Franklin Institute. Miscellaneous letters can be found in numerous other collections.

3. Benjamin Apthorp Gould, *The Ancestry and Posterity of Zaccheus Gould of Topsfield* (Salem: Printed for the Essex Institute, 1872).

4. Christa Jungnickel and Russell McCormmach, *Intellectual Mastery of Nature* (Chicago: University of Chicago Press, 1986) 65, 73–74; John Cawood, "The Magnetic Crusade: Science and Politics in Early Victorian Britain," *Isis* 70 (1979): 493–497; American Philosophical Society Memorial to Joel R. Poinsett, as quoted in *Papers of Joseph Henry*, ed. Reingold, 4:316; Bessie Zaban Jones and Lyle Gifford Boyd, *The Harvard College Observatory: The First Four Directorships, 1839–1919* (Cambridge, Mass.: Harvard University Press, 1971), 45.

5. Rosenberg, "Science and Social Values," in *Science and Values*, 22–23.

6. Gould, as quoted in George C. Comstock, "Biographical Memoir of Benjamin Apthorp Gould," *Proceedings of the National Academy of Sciences* 17 (1922): 156.

7. Newcomb, *Reminiscences of an Astronomer*, 105, 286, 288–289; Wilfrid Airy, ed., *Autobiography of Sir George Biddle Airy* (Cambridge: Cambridge University Press, 1896), v, 2–3, 212; Clerke, *Popular History of Astronomy*, 79; *Monthly Notices of the Royal Astronomical Society* 8 (1848): 211, 215; Phebe Mitchell Kendall, ed., *Maria Mitchell: Life, Letters, and Journals* (Boston: Lee and Shepard Publishers, 1896), 95–96.

8. Reingold, ed., *Papers of Joseph Henry*, 3:xviii, 375–376n

9. R. Steven Turner, "The Growth of Professorial Research in Prussia, 1818–1848—Causes and Context," in *Historical Studies in the Physical Sciences*, ed. Russell McCormmach (Philadelphia: University of Pennsylvania Press, 1971), 156–164; Jungnickel and McCormmach, *Intellectual Mastery of Nature*, 78–112; Friedrich Paulsen, *German Education Past and Present*, trans. T. Lorenz (London: T. Fisher Unwin, 1908), 184–188; Fritz K. Ringer, *The Decline of the German Mandarins: The German Academic Community, 1890–1933* (Cambridge, Mass.: Harvard University Press, 1969), 104.

10. BAG to Augustus Addison Gould, 11 April 1846, Houghton Library, Harvard University.

11. Benjamin Apthorp Gould, "On the Meridian Instruments of the Dudley Observatory," *Proceedings of the American Association for the Advancement of Science* 10 (August 1856): 113–119.

12. Marc Rothenberg, "The Educational and Intellectual Back-

ground of American Astronomers, 1825–1875." (Ph.D. diss., Bryn Mawr College, 1974), 26–27, 62.

13. Benjamin Apthorp Gould, *Report on the History of the Discovery of Neptune* (Washington City: The Smithsonian Institution, 1850).

14. According to D. B. Herrmann, Gould never studied under Gauss in Göttingen because Gauss was too busy. Herrmann argues that Gauss asked H. C. Schumacher, founder of the *Astronomische Nachrichten,* to instruct Gould. BAG to Augustus Addison Gould, 12 May 1847, Houghton Library, Harvard University; BAG to Gauss, 18 March 1847, 23 March 1847, 1 April 1847, American Philosophical Society; D. B. Herrmann, "B. A. Gould and His *Astronomical Journal,*" *Journal for the History of Astronomy* 2 (June 1971): 99.

15. BAG to Augustus Addison Gould, 12 May 1847, Houghton Library, Harvard University; Benjamin Apthorp Gould, Jr., *An Address in Commemoration of Sears Cook Walker, delivered before the American Association for the Advancement of Science, April 29, 1854* (Cambridge, Mass.: Metcalf and Company, 1854), 7.

16. BAG to Baron von Humboldt, as quoted in S. C. Chandler, "Benjamin Apthorp Gould," *Monthly Notices of the Royal Astronomical Society* 57 (February 1897): 221.

17. Gould, as quoted in Herrmann, "B. A. Gould and His *Astronomical Journal,*" 101.

18. BAG to Augustus Addison Gould, 11 April 1846, Houghton Library, Harvard University; Joseph Henry to Alexander Dallas Bache, 9 August 1838, as quoted in *Science in Nineteenth Century America,* ed. Reingold, 84; see also William Couper [to J. L. LeConte], 2 October 1854, J. L. LeConte Collection, American Philosophical Society; Donald deB. Beaver, "Altruism, Patriotism, and Science: Scientific Journals in the Early Republic," *American Studies* 12 (1971): 17.

19. Russell McCormmach, "Ormsby Macknight Mitchel's *Sidereal Messenger,* 1846–1848," *Proceedings of the American Philosophical Society* 110 (1966): 46.

20. BAG to Elias Loomis, 27 September 1849, 24 September 1850, Loomis Collection, Beinecke Library, Yale University; [B. A. Gould], "Preamble," *Astronomical Journal* 1 (1849): 1; J. S. Hubbard, "On the Establishment of An Astronomical Journal in the United States," *Summarized Proceedings of the American Association for the Advancement of Science* 2 (1849): 378–381.

21. George Bond to C. H. F. Peters, 20 August 1858, F. Brünnow to George Bond, 10 November 1858; George Bond to F. Brünnow, 27

June 1859; George Bond to J. S. Hubbard, 9 March 1859; J. S. Hubbard to George Bond, 26 April 1859, George Bond Correspondence, HUA. I am grateful to Marc Rothenberg for bringing this correspondence to my attention.

22. BAG to Airy, 11 May 1854; 23 April 1855, AC.

23. BP to ADB, 16 October 1853, PC; BAG to Elias Loomis, 4 January 1850, Loomis Collection, Beinecke Library, Yale University; BAG to ADB, 5 March [1861], RC.

24. George H. Daniels, "The Pure Science Ideal and Democratic Culture," *Science* 156 (30 June 1967): 1700.

25. BP to ADB, 21 September 1851, PC.

26. BAG to "My dear Silliman," 21 September 1860, Silliman Collection, Sterling Library, Yale University.

27. BAG to BP, 11 April 1867, PC.

28. BAG to George Biddle Airy, 11 May 1854, AC; Nathan Reingold, "Alexander Dallas Bache: Science and Technology in the American Idiom," *Technology and Culture* 2 (April 1970):170–171.

29. Bledstein, *Culture of Professionalism*, 30, 42, 114.

30. ADB to BP, 4 December 1858, PC.

31. I am deeply indebted to W. Sidman Barber, M.D., psychiatrist with Clinical Associates, Summit, New Jersey, for hours spent discussing modern psychiatry and the clinical diagnosis of manic-depressive personality disorders. Dr. Barber was extraordinarily generous with his time. I would also like to thank John Lankford for suggesting that I pursue this question.

32. Max Weber, "Science as a Vocation," in *Max Weber: Essays in Sociology*, ed., H. H. Gerth and C. W. Mills (New York: Oxford University Press, 1946), 150.

Chapter Five: An Ideal Observatory

1. Benjamin Peirce, as quoted in Blandina Dudley to TWO, 3 September 1855, Olcott Collection, Butler Library, Columbia University; ADB to TWO, 8 September 1855, 12 September 1855, OC; Gertrude E. Bell, "The Observatory-Building Movement in Nineteenth Century America," (unpublished proseminar paper, New York University, 1977), 20–21.

2. BAG to TWO, 11 September 1855, OC.

3. Mr. Simms, "On the Manufacture of Optical Glass in England,"

Monthly Notices of the Royal Astronomical Society 9 (1849): 147–148; B. K. Johnson, *Optics and Optical Instruments* (New York: Dover Publications, 1960), 159–203; Deborah Jean Warner, *Alvan Clark & Sons: Artists in Optics* (Washington, D.C.: Smithsonian Institution Press, 1968), 19–20.

4. George Biddle Airy, "Remarks," *Monthly Notices of the Royal Astronomical Society* 10 (1850): 84.

5. BAG to Elias Loomis, 29 July 1852, 1 December 1858, Loomis Collection, Beinecke Library, Yale University; Contract with Pistor and Martins of Berlin, 26 October 1855, OC.

6. BAG to TWO, 11 September 1855, OC.

7. ADB to TWO, 12 September 1855, OC.

8. BP to ADB, 25 September 1855, PC; ADB to Board of Trustees of the Dudley Observatory, 29 September 1855, as quoted in *Albany Atlas and Argus,* 30 June 1858.

9. BAG to G. B. Airy, 4 October 1855, AC; BAG to ADB, [around 16 October 1855], PC.

10. BAG to ADB, [around 16 October 1855], PC; Anthony Hyman, *Charles Babbage: Pioneer of the Computer* (Princeton, N.J.: Princeton University Press, 1982), 239–240.

11. BAG to Airy, 24 September 1856, AC.

12. G. B. Airy to Mrs. Dudley, 12 December 1855, AC.

13. Jacob Bailey, "Charles A. Spencer," *Proceedings of the American Association for the Advancement of Science* 6 (1851): 397–398; James H. Cassedy, "The Microscope in American Medical Science, 1840–1860," *Isis* 67 (March 1976): 82–87.

14. BAG to Airy, 3 January 1856, AC.

15. C. A. Spencer to TWO, 17 January 1856, OC.

16. Personal Journal of J. V. L. Pruyn, 4 January 1856, Pruyn Collection, New York State Library; BAG to Airy, 9 January 1856, AC.

17. Minutes of the Meeting of the Trustees, 11 September 1856, OC; BAG to Board of Trustees, as quoted in Dudley Observatory, *Statement of the Trustees,* 16; BAG to Airy, 24 September 1856, AC.

18. C. A. Spencer to TWO, 17 January 1856, OC.

19. Airy to BAG, 21 April 1855, BAG to Airy, 3 January 1856, BAG to Airy, 24 September 1856, AC; Proposal for a Time Ball, no date, BC; BAG to BP, 20 January 1856, PC; Munsell, *Collections* 1:471.

20. BAG to Francis Lieber, 23 April 1856, RC.

21. BAG to James Hall, 22 April 1856, Hall Collection, New York State Library; Dr. Armsby to BAG, 20 July 1856, as quoted in Gould, *Reply to the Statement of the Trustees,* 181; BAG to ADB, [around 8 August

1856], Smithsonian Institution Archives; James Hall to John Fries Frazer, 3 August 1856, J. F. Frazer Collection, American Philosophical Society.

22. BAG to Professor Challis, [no date], 1856 Correspondence Files, Cambridge Observatory Archives.

23. BAG to Armsby, 23 April 1856, 24 April 1856, 26 May 1856 as quoted in [B. A. Gould], *Specimens of the Garbling of Letters by the Majority of the Trustees of the Dudley Observatory* (Albany: Van Benthuysen, Printer, 1858), 7–9; BAG to James Hall, 22 April 1856, James Hall Collection, New York State Library; BAG to J. L. LeConte, 8 May 1856, LeConte Collection, American Philosophical Society.

24. ADB, Joseph Henry, BP to TWO, 11 August 1858, BC; Gould, *Reply to the Statement of the Trustees*, 51; BAG to ADB, 5 March 1861, RC.

25. Gould, *Reply to the Statement of the Trustees*, 52–54; Thomas Bender, "The 'Rural' Cemetery Movement: Urban Travail and the Appeal of Nature," *New England Quarterly* 47 (1974): 196–211; Goodman, "Ethics and Enterprise," 443–450.

26. Gould, *Reply to the Statement of the Trustees*, 13.

27. Remarks of Thomas Olcott to Board of Trustees, 26 June 1858, as quoted in *Albany Atlas and Argus*, 30 June 1858; Gould, *Reply to the Statement of the Trustees*, 55.

28. BAG to Armsby, 26 May 1856, as quoted in Gould, *Specimens of the Garbling of Letters*, 7–9.

29. William Rogers, as quoted in *Albany Atlas and Argus*, 21 August 1856; William Rogers to H. Rogers, 1 September 1856, as quoted in Robert V. Bruce, "Universities and the Rise of Professions: Nineteenth Century American Scientists," (Paper presented at annual meeting of the American Historical Association, December 1970), 14–15.

30. Gould, "Meridian Instruments of the Dudley Observatory," 113–119.

31. James Hall to John Fries Frazer, 3 August 1856, Frazer Collection, American Philosophical Society.

32. Dudley Observatory, *Statement of the Trustees*, 24–25.

33. BAG to F. Lieber, 25 April 1856, Huntington Library; BAG to Wolcott Gibbs, 24 November 1894, GC; Samuel Lothrop Thorndike, "Memoir of Benjamin Apthorp Gould, LL.D.," *Publications of the Colonial Society of Massachusetts* 3 (1900): 476–488.

34. Benjamin Apthorp Gould, Jr., "Remarks of Dr. Gould," in Dudley Observatory, *Inauguration of the Dudley Observatory at Albany, August 28, 1856* (Albany: Charles Van Benthuysen, Print., 1858), 37–40.

35. Rutger B. Miller to TWO, 22 May 1856, 22 May 1856, Note by Olcott, 27 June 1856, Rutger B. Miller to TWO, 27 December 1856, OC.

36. BAG to John L. LeConte, 15 September 1856, J. L. LeConte Collection, American Philosophical Society; BAG to ADB, [around 11 August 1856], Smithsonian Institution Archives.

37. BAG to BP, 16 September 1856, PC; J. D. Dana to James Hall, 5 September 1856, as quoted in Clarke, *James Hall*, 323; Albert Hopkin to William Cranch Bond, September 1856, Harvard Observatory Letterbook, Observatory Correspondence, HUA.

38. BAG to Airy, 24 September 1856, AC; Dudley Observatory, *Statement of the Trustees*, 28.

39. C. H. F. Peters to BAG, 18 September 1856, as quoted in Gould, *Reply to the Statement of the Trustees*, 322.

40. Gould, *Reply to the Statement of the Trustees*, 324.

41. Board of Trustees to Joseph Henry, 26 June 1858, BC.

42. ADB to BP, 1 November 1856, scrapbook, PC.

43. ADB to BP, 3 December 1856, scrapbook, PC.

44. Joseph Henry to BP, 28 December 1856, BAG to BP, 11 April 1867, PC; Gould, *Reply to the Statement of the Trustees*, 216–217; for conflict with Peirce over the Harvard Observatory, see Hawley to Gibbs, 3 March 1865, 8 March 1865; George W. Dean to Rev. Thomas Hill, 7 March 1865; Chauvenet to Gibbs, 9 March 1865, 14 March 1865; Martin Brimmer to B.P. Winslow, 17 March 1865, among other letters, GC.

Chapter Six: Conflict Begins: The Peters Problem

1. C. F. Winslow to John Warner, 11 November 1857, John Warner Collection, American Philosophical Society; BP to J. L. LeConte, 23 May 1858, J. L. LeConte Collection, American Philosophical Society; BP to ADB, 25 June 1857, PC.

2. *Albany Atlas and Argus*, 19 February 1857, 23 February 1857.

3. Armsby to BAG, 20 April 1857, BAG to Armsby, 21 April 1857, as quoted in Gould, *Specimens of the Garbling of Letters*, 10–11; BAG to TWO, 12 April 1858, as quoted in Benjamin Apthorp Gould, Jr., *Who Withholds Cooperation? Correspondence between the Officers of the Board of Trustees of the Dudley Observatory and the Director of the Same Institution* (Albany, N. Y.: n.p., 1858), 11–13, BAG to J. L. LeConte, 28 May 1857, J. L. LeConte Collection, American Philosophical Society.

4. C. H. F. Peters to ADB, 14 June 1857, RC.

5. C. H. F. Peters to W. C. Bond, 3 August 1857, Harvard Observatory Letterbooks, HUA; *Albany Atlas and Argus,* 31 July 1857.

6. BAG, as quoted in G. B. Hind to George Bond, 27 May 1861, George Bond Correspondence, HUA; Benjamin Apthorp Gould, *Review of "Outlines of Astronomy" by Sir John F. W. Herschel* (Cambridge, Mass.: Metcalf and Company, 1849), 15.

7. BAG to Dr. Peters, 4 August 1857, as quoted in Gould, *Specimens of the Garbling of Letters,* 14.

8. BAG to TWO, 3 October 1857, as quoted in Gould, *Specimens of the Garbling of Letters,* 17; Bache, as quoted in C. H. F. Peters to ADB, 11 November 1857, PC.

9. ADB to BP, 13 November 1857, 23 November 1857; ADB to C. H. F. Peters, 13 November 1857, PC.

10. Gould, *Reply to the Statement of the Trustees,* 57–58.

11. ADB to BP, 23 November 1857, PC.

12. ADB to BP, 26 November 1857, 30 November 1857, 3 December 1857, PC.

13. ADB to BP, 3 December 1857, PC; Gould, *Reply to the Statement of the Trustees,* 56–68.

14. ADB to BP, 6 December 1857, PC; [BAG] to "My dear Sir [Airy]," 26 November 1857, AC.

15. ADB to BP, 14 December 1857, PC.

16. ADB to BP, 21 December 1857, PC; Gould, *Reply to the Statement of the Trustees,* 59.

17. BP to ADB, 23 December 1857, RC.

18. ADB to BP, 21 December 1857, PC.

19. ADB to BP, 25 December 1857, PC.

20. BAG to BP, 1 January 1858, PC.

21. C. H. F. Peters to BAG, as quoted in Gould, *Reply to the Statement of the Trustees,* 64.

22. Personal Journal of J. V. L. Pruyn, 8 January 1858, 19 January 1858, Pruyn Collection, New York State Library; Thurlow Weed to ADB, 2 January 1858, RC; Louis Agassiz to James Hall, 4 January 1858, Hall Collection, New York State Library.

23. Copy of Minutes of Board of Trustees Meeting, 9 January 1858, BAG to BP, 1 January [1858], PC.

24. Draft of Opinion by the Scientific Council, BC.

25. BAG to BP, 10 January [1858], PC.

26. Copy of Board of Trustees Resolutions, PC.

27. For a different interpretation of the significance of the Peters appointment, see Olson, "Gould Controversy at Dudley Observatory," 273.

28. In addition to his iron business, Erastus Corning's other interests included directorships in the Albany City Bank, the New York State Bank, the United States Insurance Company, and the Merchants Insurance Company of Albany. He had a succession of partners in his own enterprises and usually exercised a leadership role in any business venture in which he was involved. He was a powerful "behind-the-scenes" politician, a member of the Albany Regency, and former mayor of Albany. Irene D. Neu, *Erastus Corning: Merchant and Financier, 1794–1872* (Ithaca, N.Y.: Cornell University Press, 1960); *DAB* 4:446–447; Reingold, ed., *Papers of Joseph Henry* 2:120n, 151n.

29. ADB to BP, 11 January 1858, BAG to BP, 14 January 1858, PC.

30. J. E. Gavit to ADB, 15 January 1858, ADB to J. E. Gavit [15 January 1858], ADB to BP and Joseph Henry, 15 January 1858, ADB to J. E. Gavit, 15 January 1858, BC; Personal Journal of J. V. L. Pruyn, 14 January 1858 through 17 January 1858, Pruyn Collection, New York State Library; Gould, *Reply to the Statement of the Trustees,* 65.

31. James Hall to Joseph Henry, 15 January 1858, Hall Collection, New York State Library; Dudley Observatory, *Statement of the Trustees,* 58–61; Scientific Council, *Defence of Dr. Gould by the Scientific Council of the Dudley Observatory* (Albany: Weed, Parsons and Company, 1858), 21.

32. Dudley Observatory, *Statement of the Trustees,* 61–62; Gould, *Reply to the Statement of the Trustees,* 67; Personal Journal of J. V. L. Pruyn, 19 January 1858, Pruyn Collection, New York State Library; Board of Trustees to Joseph Henry, 26 June 1858, BC.

33. Griffin, *Their Brothers' Keepers,* 52–60.

34. BAG to BP, 18 January 1858, PC; Scientific Council, *Defence of Dr. Gould,* 21; Gould, *Reply to the Statement of the Trustees,* 5, 57–58.

35. BP to ADB, 24 January [1858], RC.

36. ADB to BP, 27 January 1858, PC.

37. George Bond to F. Brünnow, 27 June 1859, George Bond Correspondence, HUA.

Chapter Seven: Money, Misunderstandings, and Mistrust

1. James Hall to BAG, 7 February 1858, James Hall Collection, New York State Archives; James Hall to ADB, 18 June 1858, BC.

2. BAG to the Board of Trustees, 21 January 1858, as quoted in Gould, *Reply to the Statement of the Trustees,* 279–280; BAG to TWO, 28 January 1858, as quoted in Gould, *Specimens of the Garbling of Letters,* 18.

3. William Cranch Bond, "Description of the Observatory at Cambridge, Massachusetts," *Memoirs of the American Academy of Arts and Sciences* n.s. 4 (1849): 180–181; Eufrosina Dvoichenko-Markov, "The Pulkova Observatory and Some American Astronomers of the Mid-Nineteenth Century," *Isis* 43 (1952): 244; G. W. Hough, "Description of the Buildings and Instruments," in *Annals of the Dudley Observatory* (Albany: Weed, Parsons and Company, 1866), 1:6–9; Scientific Council, *Defence of Dr. Gould,* 11; BAG to Board of Trustees, 27 January 1858, as quoted in Gould, *Who Withholds Cooperation?*, 8–9.

4. Remarks of Thomas Olcott to the Board of Trustees, as quoted in *Albany Atlas and Argus,* 30 June 1858.

5. Dudley Observatory, *Statement of the Trustees,* 69–70.

6. Gould, *Reply to the Statement of the Trustees,* 279; BAG to BP, 10 January 1858, PC.

7. Notes of meeting, 8 February 1858, BC; Joseph Henry, ADB, BP to TWO, 16 June 1858, PC; Scientific Council, *Defence of Dr. Gould,* 28.

8. Letter of Gould to *Knickerbocker,* 9 June 1858, unidentified clipping, BC; BAG to BP, 29 March 1858, PC.

9. Scientific Council, *Defence of Dr. Gould,* 29.

10. Dudley Observatory, *Statement of the Trustees,* 152.

11. Resolution, as quoted in Dudley Observatory, *Statement of the Trustees,* 68.

12. Richard Olson agrees that the scientists seem to have honestly believed they had the right to hold the institution through the power of their January agreement, even against the will of the trustees. Olson, "Gould Controversy at Dudley Observatory," 274.

13. Gould, *Reply to the Statement of the Trustees,* 75–76.

14. Benjamin A. Gould, Sr., to ADB, 25 February 1858, BC.

15. Goodman, "Ethics and Enterprise," 442, 446, 448.

16. W. R. P. [Palmer] to ADB, 29 September 1858, BC; Alexis Caswell to ADB, 8 April 1859, RC; Comstock, "Memoir of Benjamin Apthorp Gould," 160; Erving Winslow, "Sketch of Professor B. A. Gould," *Popular Science Monthly* 20 (March 1882): 684–685.

17. Board of Trustees to Joseph Henry, 26 June 1858, BC; Stepehn Van Rensselaer et al., *An Address to the Citizens of Albany, and the Donors and Friends of the Dudley Observatory, on the Recent Proceedings of the Trus-*

tees; from the Committee of Citizens Appointed at a Public Meeting held in Albany on the Thirteenth of July, 1858 (Albany: Comstock and Cassidy, Printers, 1858), 2.

18. Comstock, "Memoir of Benjamin Apthorp Gould," 163–164; BAG to Elias Loomis, 1 April 1867, Loomis Collection, Beinecke Library, Yale University; Benjamin Apthorp Gould, *Uranometria Argentina* (Buenos Aires: Paul Emile Coni, 1879), 3, 17–18, 103; Note, [no date], Draft of statement by Bache, [no date], BC.

19. ADB to BP, 14 May 1858, PC.

20. Scientific Council, *Defence of Dr. Gould,* 64–66; C. H. F. Peters to TWO, 7 August 1858, as quoted in Dudley Observatory, *Statement of the Trustees,* 164–167.

21. Scientific Council, *Defence of Dr. Gould,* 24–25.

22. For a different interpretation, see Olson, "The Gould Controversy at Dudley Observatory," 272; "Obituary: Dr. B. A. Gould," *The Observatory: A Monthly Review of Astronomy* 20 (1897): 71.

23. BAG to Board of Trustees, 4 March 1858, as quoted in Dudley Observatory, *Statement of the Trustees,* 71; Gould, *Who Withholds Cooperation?*, 12–17; Scientific Council, *Defence of Dr. Gould,* 13–16, 34–35.

24. Dudley Observatory, *Statement of the Trustees,* 136; Scientific Council, *Defence of Dr. Gould,* 12–13; Edward Singleton Holden, *Memorials of William Cranch Bond and of his son George Phillips Bond* (New York: Lemcke and Buechner, 1897), 21–22; Gould, *Reply to the Statement of the Trustees,* 125–129.

25. Gould, *Reply to the Statement of the Trustees,* 281.

26. Dudley Observatory, *Statement of the Trustees,* 72; J. H. Armsby to BAG, [around 9 March 1858], as quoted in Gould, *Who Withholds Cooperation?*, 17–18.

27. Board of Trustees to Joseph Henry, 26 June 1858, BC; TWO to Board of Trustees, as quoted in *Albany Atlas and Argus,* 30 June 1858; Gould, *Reply to the Statement of the Trustees,* 343.

28. BAG to BP, 29 March 1858, BAG to Trustees, 20 March 1858 [copy of letter], PC.

29. BAG to BP, 29 March 1858, PC.

30. Board of Trustees to Joseph Henry, 26 June 1858, BC.

31. Note, [no date], ADB to TWO, 26 April 1858, BC.

32. Board of Trustees to Joseph Henry, 26 June 1858, BC.

33. Board of Trustees to BAG, 7 April 1858, as quoted in Gould, *Who Withholds Cooperation?*, 26–27; Bache, as quoted in J. C. Spencer to Junius Hillyer, 6 October 1858, TWO to ADB, 24 April 1858, ADB to TWO, 26 April 1858, BC.

34. ADB to BAG, 22 April 1858, BAG to Trustees, 23 April 1858, TWO to ADB, 24 April 1858, ADB to TWO, 26 April 1858, BC.; TWO to BAG 23 April 1858, 28 April 1858, as quoted in Gould, *Who Withholds Cooperation?*, 29–30; Boss, *History of the Dudley Observatory,* 16.

35. George W. Dean to ADB, 17 May 1858, A. E. Winslow to Lt. E. D. Ashe, 14 June 1858, BAG to Trustees, 23 April 1858 [copy of letter], BC; George Bond to BAG, 9 July 1858 [copy of letter], PC; George Bond to F. Brünnow, 10 January 1860, 17 January 1860, 24 January 1860, 31 January 1860, 6 February 1860, F. Brünnow to George Bond, 26 January 1860, 1 February 1860, George Bond Correspondence, HUA; for reference to Gould's determination, see BAG to ADB, 5 April 1861, 6 April 1861, 13 April 1861, RC.

36. Goodman, "Ethics and Enterprise," 442; Pessen, "Social Configuration of the Antebellum City," 293.

Chapter Eight: The Final Straw

1. Gould, *Reply to the Statement of the Trustees,* 82.

2. John Gavit to ADB, 7 May 1858, BC; Dudley Observatory, *Statement of the Trustees,* 73.

3. D. E. Winslow to BAG, 20 May 1858, James Tilton to BAG, 20 May 1858, PC.

4. James Tilton to BAG, 21 May 1858, BAG to Scientific Council, 23 May 1858, PC.

5. BAG to *Albany Atlas and Argus,* 21 June 1858; Gould, *Reply to the Statement of the Trustees,* 102.

6. BAG to Scientific Council, 23 May 1858, James Tilton to BAG, 20 May 1858, James H. Toomer to BAG, 20 May 1858, D. E. Winslow to BAG, 20 May 1858, PC; John N. Wilder to Board of Trustees, as quoted in *Albany Atlas and Argus,* 19 June 1858.

7. BAG to Scientific Council, 23 May 1858, PC.

8. ADB to BP, 21 May 1858, D. E. Winslow to BAG, 21 May 1858, PC; TWO to BAG, 20 May 1858, as quoted in Gould, *Who Withholds Cooperation?*, 37–40; Gould, *Reply to the Statement of the Trustees,* 103.

9. "Statement of R. H. Pruyn," as quoted in Dudley Observatory, *Statement of the Trustees,* 75–76; D. E. Winslow to BAG, 21 May 1858, PC.

10. James Tilton to BAG, 21 May 1858, James Tilton to BAG, 22 May 1858, PC.

11. BAG to Scientific Council, 23 May 1858, PC.

12. BAG to BP, 27 May 1858, PC.

13. Dudley Observatory, *Statement of the Trustees,* 74; Scientific Council, *Defence of Dr. Gould,* 56; Gould, *Reply to the Statement of the Trustees,* 343–344.

14. Thomas Olcott to the Board of Trustees, as quoted in *Albany Atlas and Argus,* 30 June 1858.

15. Copy of Resolutions of Executive Committee of Board of Trustees, 22 May 1858, PC; Boss, *History of the Dudley Observatory,* 17; Diary of Mary Olcott, OC; John N. Wilder to Board of Trustees, as quoted in *Albany Atlas and Argus,* 19 June 1858.

16. BAG to BP, 23 May 1858, BAG to Scientific Council, 23 May 1858, BAG to BP, 23 May 1858, PC.

17. BAG to BP, 27 May 1858, ADB to BP, 30 May [1858], PC.

18. ADB to BP, 28 May [1858], 2 June 1858, PC.

19. BAG to BP, 27 May 1858, Joseph Henry and ADB to BAG, 28 May 1858, PC.

20. BAG to Executive Committee of the Dudley Observatory, 31 May 1858, as quoted in Gould, *Who Withholds Cooperation?,* 46; Scientific Council, *Defence of Dr. Gould,* 26; Gould, *Reply to the Statement of the Trustees,* 286.

21. Board of Trustees to Joseph Henry, 26 June 1858, BC; Dudley Observatory, *Statement of the Trustees,* 79–80.

22. Gould, *Reply to the Statement of the Trustees,* 285; Scientific Council, *Defence of Dr. Gould,* 42.

23. BAG to BP, 5 June 1858, PC; Personal Journal of J. V. L. Pruyn, 27 May 1858, Pruyn Collection, New York State Library.

24. Remarks of John N. Wilder to the Board of Trustees, as quoted in *Albany Atlas and Argus,* 19 June 1858.

25. Dudley Observatory, *Statement of the Trustees,* 83–84.

26. Resolutions of the Trustees, as quoted in Dudley Observatory, *Statement of the Trustees,* 87; *Albany Atlas and Argus,* 19 June 1858.

27. BAG to BP, 5 June 1858, PC.

28. TWO to ADB, 5 June 1858, PC.

Chapter Nine: Who Is in Charge Here?

1. ADB to BP, 10 May 1858, 30 May 1858, 2 June 1858, 10 June 1858, PC; William Parker Foulke to John Warner, 3 June 1858, J. P. Lesley to John Warner, 26 June 1858, John Warner Collection, American Philosophical Society; Madge E. Pickard, "Government and Science

in the United States: Historical Backgrounds," *Journal of the History of Medicine and Allied Sciences* 1 (1946): 286–289; Merle M. Odgers, *Alexander Dallas Bache, Scientist and Educator* (Philadelphia: University of Pennsylvania Press, 1947), 151–153.

2. BP to J. L. LeConte, 23 May 1858, J. L. LeConte Collection, American Philosophical Society.

3. ADB to BP, 6 June 1858, RC; ADB to BP, 7 June [1858], PC.

4. Wolcott Gibbs to ADB, 8 June 1858, BC; BAG to BP, 11 June 1858, PC; BAG to George Biddle Airy, 13 September 1858, AC; Wolcott Gibbs to BAG, 31 January 1889, 3 February 1889, BAG to "My dear old boy [Gibbs]," 8 March 1889, GC; Howard Plotkin, "Henry Tappan, Franz Brünnow, and the Founding of the Ann Arbor School of Astronomers, 1852–63," *Annals of Science* 37 (1980): 295; Marc Rothenberg, "Organization and Control: Professionals and Amateurs in American Astronomy, 1899–1918," *Social Studies of Science* 11 (1981): 307–308.

5. BAG to BP, 10 June 1858, PC.

6. James Hall to ADB, [June 1858], Hall Collection, New York State Library; Personal Journal of J. V. L. Pruyn, 2 October 1858, Pruyn Collection, New York State Library; BAG to BP, 17 May 1858, PC; Wolcott Gibbs to ADB, 13 June 1858, James Hall to ADB, 18 June 1858, BC.

7. BAG to BP, 10 June 1858, PC.

8. Note, [no date], BC; BAG to BP, 10 June 1858, PC; BAG to Airy, 13 September 1858, AC.

9. BAG to BP, 11 June 1858, PC; W. Freeman Galpin, "Reform Movements," in *History of the State of New York,* ed. Alexander C. Flick (New York: Columbia University Press, 1934) 6:271; *Albany Atlas and Argus,* 25 January 1857; BAG to John Fries Frazer, 29 May 1852, Frazer Collection, American Philosophical Society.

10. BAG to BP, 10 June 1858, BAG to BP, 11 June 1858, 18 June 1858, ADB to BP, 17 June 1858, 18 June [1858], PC.

11. Joseph Henry to ADB, 11 June 1858, Smithsonian Institution Archives.

12. Reingold, "Definitions and Speculations," in *Pursuit of Knowledge in the Early American Republic,* ed. Oleson and Brown, 35; Beer and Lewis, "Aspects of the Professionalization of Science," 764–784; Ernest Greenwood, "Attributes of a Profession," *Social Work* 2 (July 1957): 48; Harold L. Wilensky, "The Professionalization of Everyone?," *American Journal of Sociology* 70 (September 1964): 141.

13. Joseph Henry, ADB, BP to TWO, 16 June 1858, PC.

14. Personal Journal of J. V. L. Pruyn, 8 July 1858, Pruyn Collection, New York State Library; Note, [no date], BC.

15. James Hall to ADB, 18 June 1858, BC; *Albany Atlas and Argus*, 19 June 1858, 21 June 1858, 22 June 1858.

16. *New York Times*, 25 June 1858, 26 June 1858.

17. J. E. Hilgard to ADB, 23 June 1858, BC.

18. ADB to BP, 21 June [1858], ADB to BP, 24 June 1858, PC.

19. Dudley Observatory, *Statement of the Trustees*, 88.

20. Olcott's complaints included the following, not listed in order of importance: Gould's insistence on an expensive and useless trip to Europe by Spencer and Gavit; his refusal to attempt recovery from the insurors for the losses suffered in breakage of instruments during shipment; rejection of a stone pier costing $1,000 without explanation; expenditures of over $5,000 for a system of iron shutters and machinery that were too heavy for the dome; Gould's insistence that all work must stop until the dome was completely reconstructed; expenditures of over $1200 for clocks and chronographs that would not be needed until the heliometer was in place; Gould's preparations at trustees' expense but without their approval of plans for five new buildings for the observatory, estimated at $100,000 to $200,000 while he was a candidate for a professorship at Columbia University; Gould's disclaimer of responsibility for previous expenditures; his failure to set up the telescopes and refusal to answer questions about the progress of the work at the observatory; the unauthorized gratuity to Pistor and Martins; his refusal to correct the rude behavior of his assistants; Gould's arrogant letter justifying his assistants' conduct (Thomas Olcott, as quoted in *New York Times*, 3 July 1858).

21. See, for example, BAG to BP, 10 June 1858, PC.

22. Scientific Council, *Defence of Dr. Gould*, 53.

23. Board of Trustees to Joseph Henry, 26 June 1858, BC.

24. *New York Times*, 28 June 1858, 29 June 1858; *Albany Atlas and Argus*, 28 June 1858; Dudley Observatory, *Statement of the Trustees*, 89–91.

25. Scientific Council to TWO, 30 June 1858, TWO to Scientific Council, 30 June 1858, as quoted in *Albany Atlas and Argus*, 2 July 1858; Note, [no date], BC.

26. Letter and Resolutions of the Scientific Council, 30 June 1858, as quoted in *New York Times*, 3 July 1858.

27. Board of Trustees to Mrs. Dudley, as quoted in *New York Times*, 3 July 1858; Rutger B. Miller to TWO, 27 December 1858, OC; Dudley Observatory, *Statement of the Trustees*, 92–93.

28. George Bond to BAG, 9 July 1858 [copy of letter], PC.

29. BAG to George Bond, 29 June 1858, George Bond to BAG, 2 July 1858, BAG to George Bond, 6 July 1858, George Bond to BAG, 9 July 1858 [copies of letters], PC; George Bond to C. H. F. Peters, 20 August 1858, George Bond to C. A. F. Peters, 26 August 1859, George Bond Correspondence, HUA.

30. Joseph Henry to Louis Agassiz, 13 August 1864, PC; George Bond to C. A. F. Peters, 20 August 1858, George Bond Correspondence, HUA.

31. Scientific Council to Board of Trustees, 1 July 1858, TWO to Scientific Council, 1 July 1858, as quoted in *Albany Atlas and Argus,* 2 July 1858.

32. OMM to "My dear Ripley," 30 May 1859, Mitchel Collection, Cincinnati Historical Society; BAG to Airy, 11 May 1854, AC; F. W. Mitchel, *Ormsby MacKnight Mitchel: Astronomer and General* (Boston: Houghton, Mifflin Company, 1887), 190–191; Olson, "Gould Controversy at Dudley Observatory," 269.

Chapter Ten: The Pamphlet War Begins

1. ADB to Honorable Howell Cobb, 4 July 1858, W. R. Palmer to ADB, 6 July 1858 [telegram], W. R. Palmer to ADB, 6 July 1858, BC.

2. TWO to Joseph Henry, BP, BAG, ADB, 6 July 1858, Wolcott Gibbs to ADB, 7 July 1858, Theodore Sedgewick to ADB, 7 July 1858, BC.

3. ADB to James C. Spencer, 7 July 1858, BC.

4. James C. Spencer to ADB, 8 July 1858, Wolcott Gibbs to ADB, 8 July 1858, John H. Reynolds to ADB, 10 July 1858, BC; ADB to BP, 8 July 1858, PC.

5. Personal Journal of J. V. L. Pruyn, 8 July 1858, Pruyn Collection, New York State Library; Joseph Henry to ADB, 18 October 1858, Smithsonian Institution Archives.

6. *Albany Atlas and Argus,* 10 July 1858; Benjamin Apthorp Gould, Sr., to ADB, 15 July 1858, BC; ADB to BP, 17 July 1858, BAG to BP, 17 July 1858, PC.

7. Personal Journal of J. V. L. Pruyn, 12 July 1858, Pruyn Collection, New York State Library.

8. C. F. Winslow to John Warner, 7 July 1858, 11 September 1858, [John Warner] to C. F. Winslow, 16 September 1858, 31 October 1858,

John Warner Collection, American Philosophical Society; BAG to BP, 17 July 1858, PC. The *Boston Courier* listed the newspapers that supported the trustees. *Boston Courier,* 28 September 1858; *Albany Atlas and Argus,* 10 July 1858.

9. ADB to Captain Palmer, 12 July 1858, BC.

10. The twelve authors of the pamphlet included Stephen Van Rensselaer, D. D. Barnard, Erastus Corning, J. B. Plumb, James Edwards, Thomas Hun, Orlando Meads, Mason F. Cogswell, John Tayler Cooper, John V. L. Pruyn, Isaac Vanderpoel, and Harmon Pumelly. Of these, Corning, Hun, Meads, and Pumpelly were friends of Joseph Henry. Corning was also on close terms with Bache. Thomas Hun was a good friend of Benjamin Peirce. *New York Times,* 14 July 1858; Personal Journal of J. V. L. Pruyn, 13 July 1858, Pruyn Collection, New York State Library; ADB to BP, 13 July 1858, PC.

11. Van Rensselaer et al., *Address to the Citizens of Albany,* 7.

12. Jasper Adams, "On the Relation Subsisting between The Board of Trustees and Faculty of a University," as quoted in *American Higher Education: A Documentary History,* ed., Richard Hofstadter and Wilson Smith, (Chicago: University of Chicago Press, 1961), 1:320.

13. Richard Hofstadter and C. DeWitt Hardy, *The Development and Scope of Higher Education in the United States* (New York: Columbia University Press, 1952), 126–131; Story, *Forging of an Aristocracy,* 57–73; Frederick Rudolph, *The American College and University: A History* (New York: Alfred A. Knopf, 1962), 175–176.

14. Van Rensselaer et al., *Address to the Citizens of Albany,* 8, 16–17.

15. Personal Journal of J. V. L. Pruyn, 14 July 1858, Pruyn Collection, New York State Library; ADB to BP, 16 July [1858], BAG to BP, 16 July 1858, PC; W. R. P. [Palmer] to ADB, 14 July 1858, BC.

16. BAG to BP, 17 July 1858, PC; *Albany Atlas and Argus,* 16 July 1858, 22 July 1858.

17. Howell Cobb to ADB, 24 July 1858, ADB to James C. Spencer, 26 July 1858, BC; John L. LeConte to "My dear Professor [Peirce]" 19 July 1858, BAG to BP, 26 July 1858, BP to ADB, 31 July 1858, Thomas Hun to BP, 3 August 1858, PC.

18. BAG to Elias Loomis, 30 July 1858, Loomis Collection, Beinecke Library, Yale University; BAG to John L. LeConte, 6 August 1858, J. L. LeConte Collection, American Philosophical Society.

19. James C. Spencer to ADB, 9 August 1858, ADB to Spencer, 17 August 1858, BC.

20. Junius Hillyer to ADB, 10 August 1858, BC.

21. ADB to James C. Spencer, 17 August 1858, James C. Spencer to ADB, 30 August 1858, BC.

22. James Alfred Pearce to ADB, 17 August 1858, RC.

23. BAG to BP, 30 August 1858, ADB to BP, 11 October [1858], PC; William Letchworth to TWO, 22 November 1858, Olcott Collection, Butler Library, Columbia University.

24. BAG to BP, 30 August 1858, PC; see also, Gould, *Specimens of the Garbling of Letters.*

25. Joseph Henry to "My dear B.[Bache]," 31 August 1858, Smithsonian Institution Archives; Wolcott Gibbs to ADB, 7 September 1858, BC; Personal Journal of J. V. L. Pruyn, 30 August 1858, Pruyn Collection, New York State Library; ADB to BP, 21 September 1858, PC.

Chapter Eleven: War on Two Fronts

1. *Boston Advertiser,* 23 September 1858.

2. TWO to Honorable Howell Cobb, 2 September 1858, Hill, Cagger, and Porter to Howell Cobb, 2 September 1858, W. R. Palmer to ADB or BP, 6 September 1858 [telegram], B. F. Pleasants to ADB, 6 September 1858, BC.

3. ADB to Erastus Corning, 6 September 1858, ADB to Captain W. R. Palmer, 6 September 1858, BC.

4. BAG to Hon. John V. L. Pruyn, 8 September 1858, BC.

5. ADB to Capt. W. R. Palmer, 9 September 1858 [telegram], James P. Roy to ADB, 10 September 1858, Junius Hillyer to J. C. Spencer, 9 September 1858, BC.

6. James P. Roy to ADB, 10 September 1858 [telegram], ADB to Lt. James P. Roy, USN, 11 September 1858 [telegram], W. R. Palmer to ADB, 13 September 1858 [telegram], BC.

7. BAG to G. B. Airy, 13 September 1858, AC.

8. Junius Hillyer to ADB, 10 September 1858, 14 September 1858, ADB to Junius Hillyer, 18 September 1858, BC.

9. John H. Reynolds to ADB, 17 September 1858, BC.

10. Mrs. Dudley to TWO, 17 September 1858, Announcement of Propositions by Mr. Miller, 17 September 1858, Filing of Complaint, all as quoted in *New York Times,* 29 September 1858.

11. *Albany Atlas and Argus,* 20 September 1858; "Reply of the Trustees to the Charge of Garbling," *Albany Atlas and Argus,* 20 September 1858.

12. Joseph Henry to ADB, 18 September 1858, Smithsonian Institution Archives; Wolcott Gibbs to ADB, 7 September 1858, BC.

13. ADB to Junius Hillyer, 21 September 1858, BC.

14. W. R. P.[Palmer] to ADB, 29 September 1858, BC.

15. W. R. P.[Palmer] to ADB, 29 September 1858, BC.

16. Personal Journal of J. V. L. Pruyn, 2 October 1858, Pruyn Collection, New York State Library; James Hall to James D. Dana, 10 September 1858, as quoted in Clarke, *James Hall,* 350–51.

17. "Observer" [George Hornell Thacher], *A Letter to the Majority of the Trustees of the Dudley Observatory* (Albany: n.p. 1858); George H. Thacher, *A Key to the "Trustee's Statement," Letters to the Majority of the Trustees of the Dudley Observatory, showing the Misrepresentations, Garblings and Perversions of their Mis-Statement* (Albany: Atlas and Argus, 1858); *Albany Atlas and Argus,* 9 October 1858, 12 October 1858, 16 October 1858, 20 October 1858; 23 October 1858, 9 November 1858, 13 November 1858, 15 November 1858.

18. Junius Hillyer to Hon. Howell Cobb, 1 October 1858, BC.

19. Howell Cobb to ADB, 5 October 1858, BC.

20. J. C. Spencer to Hon. Junius Hillyer, 6 October 1858, Treasury Department to J. C. Spencer, 6 October 1858, BC; ADB to BP, 5 October 1858, PC.

21. ADB to BP, 5 October 1858, PC; Jefferson Davis to ADB, 22 October 1858, BC.

22. Rough draft of letter from ADB to Hon. Howell Cobb, 9 October 1858, BC.

23. ADB to BP, 18 October [1858], PC.

24. BP to ADB, 9 October 1858, RC; [Jefferson] Davis to ADB, 15 October 1858, BC.

25. TWO to Hon. Howell Cobb, 12 October 1858 [copy of letter], Howell Cobb to ADB, 14 October 1858, BC.

26. W. R. Palmer to ADB, 14 October 1858, BC.

27. Joseph Henry to ADB, 18 October 1858, Smithsonian Institution Archives.

28. BP to ADB, 23 December 1857, RC; Jefferson Davis to ADB, 22 October 1858, BC.

29. Jefferson Davis to ADB, 27 October 1858, J. E. Hilgard to ADB, 28 October 1858, BC.

30. Junius Hillyer to Hon. Howell Cobb, 23 October 1858 [copy of letter], BC.

31. Editorial from undesignated newspaper, BC.

Chapter Twelve: The Last Campaign

1. Personal Journal of J. V. L. Pruyn, 3 November 1858, Pruyn Collection, New York State Library; clipping from undesignated newspaper, BC.

2. ADB to BP, 5 November [1858], PC.

3. William P. Letchworth to TWO, 22 November 1858, Olcott Collection, Butler Library, Columbia University; BP to John Fries Frazer, 2 November 1858, Frazer Collection, American Philosophical Society.

4. BAG to Mrs. James Hall, 21 November [1858], Hall Collection, New York State Library.

5. ADB to BP, 4 December 1858, BAG to BP, 6 December 1858, PC.

6. Jefferson Davis to ADB, 22 October 1858, Petition [undated], James H. Armsby to ADB, 9 December 1858, BC; James H. Armsby to BP, 9 December 1858, PC.

7. Wolcott Gibbs to ADB, 21 December 1858, RC; BAG to BP, 21 January 1859, PC; Comstock, "Memoir of Benjamin Apthorp Gould," 159.

8. Authorization from Olcott, as quoted in *Albany Atlas and Argus,* 13 January 1859; *New York Times,* 14 January 1859.

9. BAG to BP, 4 January 1859, PC.

10. George Bond to Dr. Armsby, 20 January 1859, as quoted in Holden, *Memorials of William Cranch Bond,* 161.

11. BAG to BP, 4 January 1859, ADB to BP, 7 January [1859], PC; Gould, "To the Donors and Friends of the Dudley Observatory," *Albany Atlas and Argus,* 6 January 1859; Elisha Mack, "Statement," *Albany Atlas and Argus,* 13 January 1859.

12. Gould, "To the Donors and Friends," *Albany Atlas and Argus,* 6 January 1859.

13. *Albany Atlas and Argus,* 4 January 1859.

14. ADB to Hon. J. Black, 5 January 1859, James Black to ADB, 5 January 1859, Erastus Corning to ADB, 6 January 1859, BC; W. Chauvenet to ADB, 6 January 1859, PC; Joseph Henry to ADB, 12 January 1859, Smithsonian Institution Archives.

15. ADB to BP, 16 September 1859, PC.

16. Howard Miller has described this split between Henry and Bache over the Dudley Observatory as the crucial weakening of the Lazzaroni as a force in American science. Miller, *Dollars for Research,* 66; ADB to BP, 13 January 1859, PC.

17. ADB to Hon. Howell Cobb, 12 January 1859 [copy of letter], see also, ADB to BAG, 14 January 1859, noting receipt of Gould's report of 8 January 1859, ADB to Hon. J. A. Pearce, 14 January 1859, BC.

18. TWO to Hon. Howell Cobb, 15 January 1859, BC.

19. Howell Cobb to TWO, 24 March 1859, BC; Hough, *Annals of the Dudley Observatory,* 1:22.

20. BAG to BP, 21 January 1859, PC; BAG to John Fries Frazer, 12 January 1859, Frazer Collection, American Philosophical Society; BAG to Elias Loomis, 24 January 1859, Loomis Collection, Beinecke Library, Yale University.

21. Wolcott Gibbs to BP, 2 February 1859, PC.

22. BP to ADB, 31 January 1859, RC; ADB to BP, 5 February [1859], BAG to BP, 8 February 1859, PC; Jones and Boyd, *Harvard College Observatory,* 90–95.

23. BAG to BP, 8 February 1859, BAG to BP, 20 February 1859, ADB to BP, 9 March 1859, PC; Edward Everett to George Bond, 14 March 1859, F. Brünnow to G. P. Bond, 5 February 1859, George Bond Correspondence, HUA.

24. George Bond to F. Brünnow, 27 June 1859, George Bond Correspondence, HUA; George Bond to Dr. Armsby, 20 January 1859, as quoted in Holden, *Memorials of William Cranch Bond,* 161–162.

25. George H. Thacher to Hon. Erastus Corning, 2 February 1859 [copy of letter], BC.

26. George H. Thacher to O. M. Mitchel, 1 February 1859, Erastus Corning to ADB, 3 February 1859, George H. Thacher to Hon. Erastus Corning, 5 February 1859 [copy of letter], BC.

27. Howell Cobb to ADB, 5 February 1859, ADB to Hon. Howell Cobb, 7 February 1859 [copy of letter], Howell Cobb to ADB, 9 February 1859, BC.

28. Notice of application to change the charter, BC.

29. Erastus Corning to ADB, 7 February 1859, ADB to Hon. Erastus Corning, 9 February 1859, BC.

30. Erastus Corning to ADB, 9 February 1859, J. S. Black to Hon. Howell Cobb, 8 February 1859 [copy of letter], Howell Cobb to ADB, 9 February 1859, BC.

31. George Thacher to Erastus Corning, 14 February 1859 [copy of letter], BC.

32. William Mitchell to George Bond, 26 February 1859, George Bond Correspondence, HUA.

33. BAG to BP, 10 March 1859, PC; BAG to John Fries Frazer, 16

March 1859, Frazer Collection, American Philosophical Society; BAG to ADB, 21 March 1859 [copy of letter], BC.

34. Joseph Henry to James Hall, 7 May 1859, Hall Collection, New York State Archives; David A. Wells to John Warner, 22 March 1859, John Warner Collection, American Philosophical Society.

Chapter Thirteen: The Last Word

1. OMM to ADB, 5 May 1859, RC; OMM to the trustees, 1859, as quoted by Mitchel, *Ormsby Macknight Mitchel,* 193; BAG to BP, 10 March 1859, PC.

2. OMM to "My dear Ripley," 30 May 1859, Mitchel Collection, Cincinnati Historical Society.

3. F. Brünnow to G. P. Bond, 13 May 1859, 2 June 1859, 10 December 1859, George Bond to F. Brünnow, 13 May 1859, 27 June 1859, George Bond Correspondence, HUA; Rothenberg, "Organization and Control," 307–308.

4. OMM to G. B. Airy, 22 December 1859, AC; Hough, *Annals of the Dudley Observatory,* 1:4, 9–11; Plotkin, "Tappan, Brünnow, and the Ann Arbor School of Astronomers," 295.

5. Henry C. King, *The History of the Telescope* (London: Charles Griffin and Company, 1955), 242–244; *New York Times,* 21 April 1860; OMM to TWO, 6 May 1861, OC.

6. OMM to "Dear Ned," 9 May 1860, 23 March 1861, Mitchel Collection, Cincinnati Historical Society; BAG to BP, 23 July 1860, PC; Mitchel, *Ormsby Macknight Mitchel,* 202–203.

7. O. M. Mitchel to Ira Harris, 4 May 1861, RC.

8. For a description of the "Great Locomotive Chase" by a surviving participant, see pamphlet in Mitchel Collection, Cincinnati Historical Society; BAG to ADB, 13 June 1862, RC; see also, Mitchel, *Ormsby Macknight Mitchel,* 207–208.

9. J. E. Nourse, "Observatories in the United States," *Harper's New Monthly Magazine* 48 (March 1874): 526; Act of Incorporation, 10 April 1873, OC; Deborah Jean Warner, "Lewis Boss," *DSB* 2:332–333.

10. Ralph E. Wilson, "The General Catalogue of Stellar Positions and Proper Motions," *Astronomical Society of the Pacific* 3 (October 1939): 215–222; Boss, *History of the Dudley Observatory,* 47–54.

11. BAG to Wolcott Gibbs, 7 April 1891, GC; Boss, *History of the Dudley Observatory,* 38–39.

12. Gould's extensive collection of rare astronomical books was recently catalogued by Jan Ludwig, chairman of the philosophy department of Union College. The collection is now housed at Union College.

13. James Hall to BAG, 27 July 1859 [copy of letter], Hall Collection, New York State Library; ADB to BP, 20 April 1860, PC; Election tabulations, Ira Harris to TWO, 7 March 1863, OC; Dickerson to TWO, 17 November 1860, TWO to Salmon P. Chase, 17 September 1861, Olcott Collection, Butler Library, Columbia University.

14. BAG to ADB, 4 February [1861], RC.

15. BAG to "My dear Silliman," 21 September 1860, Silliman Collection, Sterling Library, Yale University; ADB to BP, 1 July 1860, PC.

16. ADB to BP, 5 July 1861, PC.

Chapter Fourteen: Reflections and Conclusions

1. Benjamin Apthorp Gould, "Address in Commemoration of Alexander Dallas Bache," *Proceedings of the American Association for the Advancement of Science* 17 (August 1868): 41–42; Comstock, "Memoir of Benjamin Apthorp Gould," 158–159; Odgers, *Alexander Dallas Bache,* 192; Dupree, *Science in the Federal Government,* 119; Miller, *Dollars for Research,* 33–47; Storr, *Beginnings of Graduate Education,* 82–93.

2. Dr. Abraham Flexner and Frederick T. Gates, as quoted in Allan Nevins, *Study in Power: John D. Rockefeller, Industrialist and Philanthropist* (New York: Charles Scribner's Sons, 1953), 2:308–309; E. Richard Brown, *Rockefeller Medicine Men: Medicine and Capitalism in America* (Berkeley: University of California Press, 1979), 10–11.

3. Terence J. Johnson, *Professions and Power* (London: The Macmillan Press, 1972), 9–10; Talcott Parsons, "Professions," in *International Encyclopedia of the Social Sciences,* ed. David L. Sills (New York: Macmillan and Free Press, 1968), 12: 545.

4. Jethro K. Lieberman, *The Tyranny of the Experts: How Professionals Are Closing the Open Society* (New York: Walker Press, 1970); Haskell, "Introduction," in *Authority of Experts,* ed. Haskell, ix–xxxix.

5. Joseph Henry to ADB, 11 June 1858, Smithsonian Institution Archives.

6. Minutes of the Board of Control, 18 February 1854, 23 February 1854, Cincinnati Astronomical Society, Cincinnati Historical Society; Freeman Dyson, "Astronomy in a Private Sphere," *Minerva* 53 (Spring 1984): 172.

7. William J. Goode, "The Theoretical Limits of Professionalization," in *The Semi-Professions and their Organization,* ed. Amitai Etzioni (New York: Free Press, 1969), 266–313; Johnson, *Professions and Power,* 23; Magali Sarfatti Larson, *The Rise of Professionalism: A Sociological Analysis* (Berkeley: University of California Press, 1977).

8. Lankford, "Amateurs versus Professionals," 11.

9. Merle Curti, *Philanthropy in the Shaping of American Higher Education* (New Brunswick, N.J.: Rutgers University Press, 1965); Bremner, *American Philanthropy.*

Bibliography

Manuscript Collections

Letters of George Biddle Airy, Royal Greenwich Observatory Archives, Herstmonceux Castle, Hailsham, East Sussex, England.

Letters and Papers of Alexander Dallas Bache, Library of Congress.

Letters and Papers of William Cranch Bond and George Phillips Bond, Observatory Correspondence, Harvard University Archives.

G. H. Brush Collection, Sterling Library, Yale University.

Cambridge Observatory Archives, University of Cambridge, England, Correspondence Files, 1849–1878.

Cincinnati Astronomical Society Minutes, 1842–1872, Cincinnati Historical Society, Cincinnati, Ohio.

James Dwight Dana Collection, Sterling Library, Yale University.

John Fries Frazer Collection, American Philosophical Society.

Oliver Wolcott Gibbs Collection, Franklin Institute, Philadelphia.

James Hall Collection, New York State Archives, Albany, New York.

James Hall Collection, New York State Library, Albany, New York.

Letters of Joseph Henry, Smithsonian Institution Archives.

Thomas Hun Papers, Albany Institute of History and Art, Albany, New York.

Faculty Records of the Lawrence Scientific School, Harvard University Archives.
John L. LeConte Collection, American Philosophical Society.
Elias Loomis Collection, Beinecke Library, Yale University.
O. C. Marsh Collection, Sterling Library, Yale University.
O. M. Mitchel Collection, Cincinnati Historical Society, Cincinnati, Ohio.
J. B. Murray Collection, Sterling Library, Yale University.
Thomas Olcott Collection, Albany Institute of History and Art, Albany, New York.
Thomas Olcott Collection, Butler Library, Columbia University.
Benjamin Peirce Collection, Houghton Library, Harvard University.
John V. L. Pruyn Collection, New York State Library.
William Jones Rhees Collection, Huntington Library, San Marino, California.
Benjamin Silliman Collection, Sterling Library, Yale University.
Henry Tappan file, Michigan Historical Collections, Bentley Historical Library, University of Michigan.
John Warner Collection, American Philosophical Society.
Woolsey Family Papers, Sterling Library, Yale University.

Interviews and Correspondence

Correspondence with Dr. Nathan Reingold, Senior Historian of the Smithsonian Institution, June–October 1985.
Conversations with W. Sidman Barber, M.D., clinical psychiatrist with Clinical Associates, Summit, New Jersey, Summer 1986.

Books, Periodicals, and Unpublished Sources

Airy, Wilfrid, ed. *Autobiography of Sir George Biddle Airy*. Cambridge: Cambridge University Press, 1896.
Airy, George Biddle. "On the Method of Observing and Recording Transits, lately introduced in America." *Philosophical Magazine and Journal of Science* 3d ser. 36 (1850): 142–150.
Airy, George Biddle. "Remarks." *Monthly Notices of the Royal Astronomical Society* 10 (1850): 105.
Albany Atlas and Argus. 1850–1860.

Albany, New York, Committee on the Celebration of the Two Hundred Fiftieth Anniversary of the Granting of the Dongan Charter. *Albany A Cradle of America.* Albany, N.Y.: Argus Company, 1936.

Aldrich, Michele Alexis L. "New York Natural History Survey, 1836–1845." Ph.D. diss., University of Texas, 1974.

Astronomical Journal. 1849–1860.

Astronomische Nachrichten. 1831; 1838; 1858–1859.

Bache, Alexander Dallas. "Remarks." *Proceedings of the American Association for the Advancement of Science* 6 (1852): xli–lx.

Bailey, Jacob. "Charles A. Spencer." *Proceedings of the American Association for the Advancement of Science* 6 (1851): 397–398.

Baltzell, E. Digby. *Philadelphia Gentlemen: The Making of a National Upper Class.* Philadelphia: University of Pennsylvania Press, 1979.

Beaver, Donald deB. "Altruism, Patriotism, and Science: Scientific Journals in the Early Republic." *American Studies* 12 (1971): 5–19.

Beer, John J., and W. David Lewis. "Aspects of the Professionalization of Science." *Daedalus* 92 (Fall 1963): 764–784.

Bell, Trudy [Gertrude] E. "'Garblings and Perversions': The Dudley Observatory Controversy." *Griffith Observer* 38 (November 1974): 2–9.

——— "The Observatory-Building Movement in Nineteenth Century America." Paper for Proseminar: U.S. History 1777–1900," New York University, 1977.

Bender, Thomas. "The 'Rural' Cemetery Movement: Urban Travail and the Appeal of Nature." *New England Quarterly* 47 (1974): 196–211.

——— "Science and the Culture of American Communities: The Nineteenth Century." *History of Education Quarterly* 16 (Spring 1976): 63–77.

"Benefactors of Education and Science." *American Journal of Education,* 2 (1856): 593–604.

Bledstein, Burton J. *The Culture of Professionalism: The Middle Class and the Development of Higher Education in America.* New York: W. W. Norton and Company, 1976.

Bond, William Cranch. "Description of the Observatory at Cambridge, Massachusetts." *Memoirs of the American Academy of Arts and Sciences* n.s. 4 (1849): 177–188.

Boss, Benjamin. *History of the Dudley Observatory, 1852–1956.* Albany: privately printed.

Boston Advertiser. 1858.

Bremner, Robert H. *American Philanthropy*. Chicago: University of Chicago Press, 1960.

Brown, E. Richard. *Rockefeller Medicine Men: Medicine and Capitalism in America*. Berkeley: University of California Press, 1979.

Bruce, Robert V. "Universities and the Rise of Professions: Nineteenth Century American Scientists." Paper presented at Annual Meeting of American Historical Association, December, 1970.

Cassedy, James H. "The Microscope in American Medical Science, 1840–1860." *Isis* 67 (1976): 76–97.

Cawood, John. "The Magnetic Crusade: Science and Politics in Early Victorian Britain." *Isis* 70 (1979): 493–518.

Chandler, S. C. "Benjamin Apthorp Gould." *Monthly Notices of the Royal Astronomical Society* 57 (February 1897): 218–222.

Christman, Henry. *Tin Horns and Calico.* New York: Henry Holt and Company, 1945.

Clarke, John M. *James Hall of Albany: Geologist and Paleontologist, 1811–1898*. Albany: n.p., 1921.

Clerke, Agnes M. *A Popular History of Astronomy during the Nineteenth Century*. London: Adam and Charles Black, 1908.

Comstock, George C. "Biographical Memoir of Benjamin Apthorp Gould." *Proceedings of the National Academy of Sciences* 17 (1922): 155–180.

Curti, Merle, *The Growth of American Thought.* New York: Harper and Brothers, 1951.

——— *Philanthropy in the Shaping of American Higher Education.* New Brunswick, New Jersey: Rutgers University Press, 1965.

Dahl, Robert A. *Who Governs? Democracy and Power in an American City.* New Haven: Yale University Press, 1961.

Daniels, George H., ed. *Nineteenth Century American Science: A Reappraisal,* Evanston, Illinois: Northwestern University Press, 1972.

——— "The Process of Professionalization in American Science: The Emergent Period, 1820–60." *Isis* 58 (Summer 1967): 151–166.

——— "The Pure Science Ideal and Democratic Culture." *Science* 156 (30 June 1967): 1699–1705.

Dudley Observatory. *The Dudley Observatory and the Scientific Council: Statement of the Trustees.* Albany: Van Benthuysen, Printer, 1858.

——— *Inauguration of the Dudley Observatory at Albany, August 28, 1856.* Albany: Charles Van Benthuysen, 1858.

Dupree, A. Hunter. *Asa Gray.* New York: Atheneum, 1968.

——— "Central Scientific Organization in the United States Government." *Minerva* 4 (Summer 1963): 453–69.

——— *Science in the Federal Government.* Cambridge, Mass.: Harvard University Press, 1957.

Dvoichenko-Markov, Eufrosina. "The Pulkovo Observatory and Some American Astronomers of the Mid-Nineteenth Century." *Isis* 43 (1952): 243–246.

Dyson, Freeman. "Astronomy in a Private Sphere." *Minerva* 53 (Spring 1984): 169–182.

Etzioni, Amitai, ed. *The Semi-Professions and their Organization.* New York: Free Press, 1969.

Fairchild, Herman L. "The History of the American Association for the Advancement of Science." *Science* 59 (25 April 1924), 365–369; (2 May 1924): 385–390, 410–415.

Flick, Alexander C. *History of the State of New York.* 10 vols. New York: Columbia University Press, 1934.

Geison, Gerald L., ed. *Professions and Professional Ideologies in America,* Chapel Hill: University of North Carolina Press, 1983.

Gerth, H. H., and C. W. Mills, eds. *Max Weber: Essays in Sociology.* New York: Oxford University Press, 1946.

Gillispie, C. C., ed. *Dictionary of Scientific Biography.* 16 vols. New York: Charles Scribner's Sons, 1974.

Goodman, Paul. "Ethics and Enterprise: The Values of a Boston Elite, 1800–1860." *American Quarterly* 18 (Fall 1966): 437–451.

Gould, Benjamin Apthorp. "Address in Commemoration of Alexander Dallas Bache." *Proceedings of the American Association for the Advancement of Science* 17 (August 1868): 1–47.

——— *The Ancestry and Posterity of Zaccheus Gould of Topsfield.* Salem: Printed for the Essex Institute, 1872.

——— *An Address in Commemoration of Sears Cook Walker, delivered before the American Association for the Advancement of Science, April 29, 1854.* Cambridge, Mass.: Metcalf and Company, 1854.

——— "On the Meridian Instruments of the Dudley Observatory." *Proceedings of the American Association for the Advancement of Science* 10 (August 1856): 113–119.

——— *Report on the History of the Discovery of Neptune.* Washington City: The Smithsonian Institution, 1850.

——— *Reply to the "Statement of the Trustees" of the Dudley Observatory.* Albany: Charles Van Benthuysen, 1859.

——— *Review of "Outlines of Astronomy" by Sir John F. W. Herschel.* Cambridge, Mass.: Metcalf and Company, 1849.

——— *Specimens of the Garbling of Letters by the Majority of the Trustees of the Dudley Observatory.* Albany: Van Benthuysen, Printer, 1858.

——— *Uranometria Argentina.* Buenos Aires: Paul Emile Coni, 1879.

——— *Who Withholds the Cooperation? Correspondence between the Officers of the Board of Trustees of the Dudley Observatory and the Director of the Same Institution.* Albany, N.Y.: n.p., 1858.

Greenwood, Ernest. "Attributes of a Profession." *Social Work* 2 (July 1957): 45–55.

Griffin, Clifford S. *Their Brothers' Keepers: Moral Stewardship in the United States, 1800–1865.* New Brunswick, N.J.: Rutgers University Press, 1960.

Hahn, Roger. *The Anatomy of a Scientific Institution: The Paris Academy of Sciences, 1666–1803.* Berkeley: University of California Press, 1971.

Hall, James. "The New York Geological Survey." *Popular Science Monthly* 22 (1883): 815–825.

Hall, Peter Dobkin. *The Organization of American Culture, 1700–1900: Private Institutions, Elites, and the Origins of American Nationality.* New York: New York University Press, 1984.

Haskell, Thomas L., ed. *The Authority of Experts: Studies in History and Theory.* Bloomington: Indiana University Press, 1984.

Hendrickson, Walter B. "Nineteenth Century State Geological Surveys: Early Government Support of Science." *Isis* 52 (September 1961): 326–340.

Herrmann, D. B. "B. A. Gould and His *Astronomical Journal.*" *Journal for the History of Astronomy* 2 (June 1971): 98–108.

Herschel, John F. W. *Outlines of Astronomy.* London: Longman, Brown, Green, and Longman, 1851.

Hofstadter, Richard, and C. DeWitt Hardy. *The Development and Scope of Higher Education in the United States.* New York: Columbia University Press, 1952.

Hofstadter, Richard, and Wilson Smith, eds. *American Higher Education: A Documentary History.* 2 vols. Chicago: University of Chicago Press, 1961.

Holden, Edward Singleton. *Memorials of William Cranch Bond and his son George Phillips Bond.* New York: Lemcke and Buechner, 1897.

Holmfeld, John D. "From Amateurs to Professionals in American Science: The Controversy Over the Proceedings of an 1853 Scientific Meeting." *Proceedings of the American Philosophical Society* 114 (February 1970): 22–35.

Hough, G. W. *Annals of the Dudley Observatory.* 2 vols. Albany: Weed, Parsons and Company, Printers, 1866.

Howell, George R., and Jonathan Tenney, eds. *History of the County of Albany, New York, from 1609 to 1886.* 3 vols. New York: W. W. Munsell and Co., Publishers, 1886.

Hubbard, J. S. "On the Establishment of an Astronomical Journal in the United States." *Summarized Proceedings of the American Association for the Advancement of Science* 2 (1849): 378–381.

Hyman, Anthony. *Charles Babbage: Pioneer of the Computer.* Princeton, N.J.: Princeton University Press, 1982.

Jaher, Frederic Cople. "Nineteenth Century Elites in Boston and New York." *Journal of Social History* 6 (Fall 1972): 32–77.

James, Henry. *Charles W. Eliot: President of Harvard University 1869–1909.* 2 vols. Boston: Houghton, Mifflin Company, 1930.

Johnson, B. K. *Optics and Optical Instruments.* New York: Dover Publications, 1960.

Johnson, Terence J. *Professions and Power.* London: The Macmillan Press, 1972.

Jones, Bessie Zaban and Lyle Gifford Boyd. *The Harvard College Observatory: The First Four Directorships, 1839–1919.* Cambridge, Mass.: Harvard University Press, 1971.

Jungnickel, Christa, and Russell McCormmach. *Intellectual Mastery of Nature. Theoretical Physics from Ohm to Einstein.* Chicago: University of Chicago Press, 1985.

Kendall, Phebe Mitchell, ed. *Maria Mitchell: Life, Letters and Journals.* Boston: Lee and Shepard Publishers, 1896.

Kennedy, William. *O Albany!* New York: Viking Penguin, 1983.

King, Henry C. *The History of the Telescope.* London: Charles Griffin and Company, 1955.

Kohlstedt, Sally Gregory. *The Formation of the American Scientific Community: The American Association for the Advancement of Science 1848–1860.* Urbana: University of Illinois Press, 1976.

Lankford, John. "Amateurs versus Professionals: The Controversy over Telescope Size in Late Victorian Science." *Isis* 72 (March 1981): 11–28.

Larson, Magali Sarfatti. *The Rise of Professionalism: A Sociological Analysis.* Berkeley: University of California Press, 1977.

Lieberman, Jethro K. *The Tyranny of the Experts: How Professionals Are Closing the Open Society.* New York: Walker Press, 1970.

Lurie, Edward. "Nineteenth Century American Science: Insights from Four Manuscripts." *Rockefeller Institute Review* 2 (February 1964): 11–19.

Lyell, Charles. *Travels in North America.* London: John Murray, 1845.
Lynn, Kenneth S., and the editors of *Daedalus. The Professions in America.* Boston: Beacon Press, 1965.
McCormmach, Russell, ed. *Historical Studies in the Physical Sciences.* Philadelphia: University of Pennsylvania Press, 1971.
——— "Ormsby Macknight Mitchel's *Sidereal Messenger,* 1846–48." *Proceedings of the American Philosophical Society* 110 (1966): 35–47.
Malone, Dumas, ed. *Dictionary of American Biography.* New York: Charles Scribner's Sons, 1934.
Merrill, George P. *The First One Hundred Years of American Geology.* New York: Hafner Publishing Company, 1964.
Miles, Wyndham D. "Public Lectures on Chemistry in the United States." *Ambix* 15 (October 1968): 129–153.
Miller, Howard S. *Dollars for Research: Science and Its Patrons in Nineteenth Century America.* Seattle: University of Washington Press, 1970.
Mitchel, F. W. *Ormsby Macknight Mitchel: Astronomer and General.* Boston: Houghton, Mifflin Company, 1887.
Monthly Notices of the Royal Astronomical Society. 1845–60.
Munsell, Joel. *The Annals of Albany.* 10 vols. Albany: J. Munsell, 1850–1860.
——— *Collections on the History of Albany.* 4 vols. Albany, N.Y.: J. Munsell, 1865.
Neu, Irene D. *Erastus Corning: Merchant and Financier, 1794–1872.* Ithaca, N.Y.: Cornell University Press, 1960.
Nevins, Allan. *Study in Power: John D. Rockefeller, Industrialist and Philanthropist.* 2 vols. New York: Charles Scribner's Sons, 1953.
Newcomb, Simon. *The Reminiscences of an Astronomer.* Boston: Houghton, Mifflin and Company, 1903.
New York State. Office of the Secretary of State, Hon. Joel T. Headley. *Census of the State of New York for 1855; Taken in Pursuance of Article Third of the Constitution of the State, and of Chapter Sixty Four of the Laws of 1855, Prepared from the Original Returns, under the Direction of Hon. Joel T. Headley, Secretary of State, by Franklin B. Hough, Superintendent of the Census.* Albany: Charles Van Benthuysen, 1857.
New York Times. 1850–61.
Nourse, J. E. "Observatories in the United States." *Harper's New Monthly Magazine* 48 (March 1874): 526–541; (July 1874): 518–531.
"Obituary: Dr. B. A. Gould." *The Observatory: A Monthly Review of Astronomy* 20 (1897): 70–72.

"Observer" [George Hornell Thacher]. *A Letter to the Majority of the Trustees of the Dudley Observatory.* Albany, n.p., 1858.

Odgers, Merle M. *Alexander Dallas Bache: Scientist and Educator.* Philadelphia: University of Pennsylvania Press, 1947.

Oleson, Alexandra, and Sanborn C. Brown, eds. *The Pursuit of Knowledge in the Early American Republic.* Baltimore: John Hopkins University Press, 1977.

Olson, Richard G. "The Gould Controversy at Dudley Observatory: Public and Professional Values in Conflict." *Annals of Science* 27 (September 1971): 265–276.

Paulsen, Friedrich. *German Education Past and Present.* Translated by T. Lorenz. London: T. Fisher Unwin, 1908.

Parsons, Talcott. "Professions." In *International Encyclopedia of the Social Sciences.* Edited by David L. Sills. New York: MacMillan and Free Press, 1968. 12: 536–546.

Pessen, Edward. *Riches, Class, and Power Before the Civil War.* Lexington, Massachusetts: D. C. Heath and Company, 1973.

——— "The Social Configuration of the Antebellum City: An Historical and Theoretical Inquiry." *Journal of Urban History* 2 (May 1976): 267–306.

Pickard, Madge E. "Government and Science in the United States: Historical Backgrounds." *Journal of the History of Medicine and Allied Sciences* 1 (1946): 254–289, 446–481.

Plotkin, Howard. "Henry Tappan, Franz Brünnow, and the Founding of the Ann Arbor School of Astronomers, 1852–63." *Annals of Science* 37 (1980): 287–302.

Reingold, Nathan. "Alexander Dallas Bache: Science and Technology in the American Idiom." *Technology and Culture* 2 (April 1970): 163–177.

Reingold, Nathan, ed. *The Papers of Joseph Henry.* 5 vols. to date. Washington, D.C.: Smithsonian Institution Press, 1972–1985.

——— *Science in Nineteenth Century America: A Documentary History.* New York: Hill and Wang, 1964.

——— *The Sciences in the American Context: New Perspectives.* Washington, D.C.: Smithsonian Institution Press, 1979.

Reynolds, Cuyler, ed. *Hudson-Mohawk Genealogical and Family Memoirs.* New York: Lewis Historical Publishing Company, 1911.

Rezneck, Samuel. *Education for a Technological Society: A Sesquicentennial History of Rensselaer Polytechnic Institute.* Troy, N.Y.: Rensselaer Polytechnic Institute, 1968.

——— "The Emergence of a Scientific Community in New York State a Century Ago." *New York History* 63 (July 1962): 211–238.

——— "A Travelling School of Science on the Erie Canal in 1826." *New York History* 40 (July 1959): 255–269.

Ringer, Fritz K., *The Decline of the German Mandarins; The German Academic Community, 1890–1933*. Cambridge, Mass.; Harvard University Press, 1969.

Rogers, Emma, ed. *Life and Letters of William Barton Rogers*. 2 vols. Boston: Houghton, Mifflin and Company, 1896.

Rossiter, Margaret. *The Emergence of Agricultural Science: Justus Liebig and the Americans, 1840–1880*. New Haven: Yale University Press, 1975.

Rothenberg, Marc. "The Educational and Intellectual Background of American Astronomers, 1825–1875." Ph.D. diss., Bryn Mawr College, 1974.

——— "Organization and Control: Professionals and Amateurs in American Astronomy, 1899–1918." *Social Studies of Science* 11 (1981): 305–325.

Schneer, Cecil J. "The Great Taconic Controversy." *Isis* 69 (June 1978): 173–191.

Schoenberg, Wallace Kenneth. "The Young Men's Association, 1833–1876: The History of a Social-Cultural Organization." Ph.D. diss., New York University, 1962.

Scientific Council. *Defence of Dr. Gould by the Scientific Council of the Dudley Observatory*. Albany: Weed, Parsons and Company, 1858.

Silliman, Benjamin, Jr. "Science in America." *New York Quarterly* 2 (1853): 443–465.

Silverman, Robert, and Mark Beach. "A National University for Upstate New York." *American Quarterly* 22 (Fall 1970): 701–713.

Simms, Mr. "On the Manufacture of Optical Glass in England." *Monthly Notices of the Royal Astronomical Society* 9 (1849): 147–148.

Stebbins, Robert A. "Avocational Science: The Amateur Routine in Archaeology and Astronomy." *International Journal of Comparative Sociology* 21 (March–June 1980): 34–48.

Storr, Richard J. *The Beginning of the Future: A Historical Approach to Graduate Education in the Arts and Sciences*. New York: McGraw-Hill Book Company, 1973.

——— *The Beginnings of Graduate Education in America*. Chicago: University of Chicago Press, 1953.

Story, Ronald. *The Forging of an Aristocracy: Harvard and the Boston Upper*

Class 1800–1870. Middletown, Conn.: Wesleyan University Press, 1980.

Thacher, George Hornell. *A Key to the "Trustees Statement," Letters to the Majority of the Trustees of the Dudley Observatory Showing the Misrepresentations, Garblings and Perversions of their Mis-Statement.* Albany: Atlas and Argus, 1858.

Thackray, Arnold, and Everett Mendelsohn, eds. *Science and Values: Patterns of Tradition and Change.* New York: Humanities Press, 1974.

Thorndike, Samuel Lothrop. "Memoir of Benjamin Apthorp Gould, LL.D." *Publications of the Colonial Society of Massachusetts* 3 (1900): 476–488.

Trollope, Frances. *Domestic Manners of the Americans.* New York: Alfred A. Knopf, 1949.

Van Rensselaer, Stephen et al. *An Address to the Citizens of Albany, and the Donors and Friends of the Dudley Observatory, on the Recent Proceedings of the Trustees; from the Committee of Citizens Appointed at a Public Meeting held in Albany on the Thirteenth of July, 1858.* Albany: Comstock and Cassidy, Printers, 1858.

Veysey, Laurence. "Who's a Professional? Who Cares?" *Reviews in American History* 3 (December 1975): 419–423.

Warner, Deborah Jean. *Alvan Clark & Sons: Artists in Optics.* Washington, D.C.: Smithsonian Institution Press, 1968.

Weise, Arthur James. *The History of the City of Albany, New York, From the Discovery of the Great River in 1524, by Verrazzano, to the Present Time.* Albany: E. H. Bender, 1884.

Wilensky, Harold L. "The Professionalization of Everyone?" *American Journal of Sociology* 70 (September 1964): 137–158.

Wilson, Ralph E. "The General Catalogue of Stellar Positions and Proper Motions." *Astronomical Society of the Pacific* 3 (October 1939): 215–222.

Winslow, Erving. "Sketch of Professor B. A. Gould." *Popular Science Monthly* 20 (March 1882): 683–687.

Zochert, Donald. "Science and the Common Man in Ante-Bellum America." *Isis* 65 (December 1974): 448–473.

Name Index

Subject Index